Engineers encounter difficult ethical problems in their practice and in research. In many ways, these problems are like design problems: They are complex and often ill-defined; resolving them involves an iterative process of analysis and synthesis; and there can be more than one acceptable solution.

This book offers a real-world, problem-centered approach to engineering ethics, using a rich collection of open-ended scenarios and case studies to develop skill in recognizing and addressing ethical issues. The book is designed to be used with active learning classroom exercises and makes extensive use of the resources on the WWW Ethics Center for Engineering and Science, http://ethics.cwru.edu.

ETHICS IN
ENGINEERING
PRACTICE AND
RESEARCH

ETHICS IN ENGINEERING PRACTICE AND RESEARCH

Caroline Whitbeck

Case Western Reserve University

CAMBRIDGE
UNIVERSITY PRESS

PUBLISHED BY THE PRESS SYNDICATE OF THE UNIVERSITY OF CAMBRIDGE
The Pitt Building, Trumpington Street, Cambridge CB2 1RP, United Kingdom

CAMBRIDGE UNIVERSITY PRESS
The Edinburgh Building, Cambridge CB2 2RU, United Kingdom
40 West 20th Street, New York, NY 10011-4211, USA
10 Stamford Road, Oakleigh, Melbourne 3166, Australia

© Cambridge University Press 1998

First published 1998

Printed in the United States of America

Typeset in New Baskerville

Library of Congress Cataloging-in-Publication Data
Whitbeck, Caroline.
Ethics in engineering practice and research / Caroline Whitbeck.
p.   cm.
Includes bibliographical references.
ISBN 0-521-47411-6
1. Engineering ethics.   I. Title.
TA157.W47   1998
174'.962 – dc21   97-18010
                                               CIP

*A catalog record for this book is available from
the British Library.*

ISBN 0 521 47411 6   hardback
ISBN 0 521 47944 4   paperback

This book is dedicated to the memory of James R. Melcher
(1936–1991)

CONTENTS

PREFACE

Ethics in Engineering Practice and Research is about professional responsibilities of engineers and applied scientists. It is about professional responsibilities: the character of problem situations in which those responsibilities must be fulfilled and the moral skills for fulfilling them. Interspersed throughout the text are open-ended scenarios that present ethically significant situations of the sort engineers and applied scientists commonly encounter. These have been set apart in centered boxes to aid the use of them in group discussion and for homework assignments. Also set apart from the text, in boxes, are fine points, which may enhance the reader's understanding but which are not essential to the main argument. Most of these fine points concern philosophical issues.

OUTLINE AND SUMMARY

The introduction on concepts provides a clarification of many general ethical terms and provides a general framework for considering ethical questions. This framework draws on readers' prior experience of moral life and of moral reflection. Other more specialized ethical concepts are introduced as needed throughout the book.

Chapter 1 discusses what moral problems look like to a person in the situation who must respond to them. The frequent need to cope with an ambiguous situation and to formulate responses to the problem situation shows that addressing ethically significant problems is more demanding than simply evaluating the relative merits of preestablished responses. In many respects challenging ethical problems resembles challenging design problems.

Chapter 2 discusses professional responsibility, its basis and scope, and provides comparison of engineering with other professions. (Beginning with this one, the order of the chapters roughly corresponds to the sophistication of their subject matter.)

The Central Professional Responsibilities of Engineers and applied scientists, especially the responsibility for safety, is the subject of Chapter 3. Public safety, consumer safety, operator safety, occupational safety, and laboratory safety are considered.

Chapter 4 recounts the stories of two engineers who discharged their responsibility for safety in exemplary ways. Their stories are told in detail to show the development of the problem situation they faced and the appropriate responses that they made at different stages.

Chapter 5 treats workplace rights and responsibilities, focusing on engineers in corporations or governmental organizations.

Chapter 6 on the responsibility for research integrity and later chapters on research ethics carry over the discussion of complaint handling in Chapter 5 to universities dealing with charges of research misconduct.

Chapter 7 examines investigator's responsibilities for the subjects of their research experiments.

Responsibility for the environment, which is the subject of Chapter 8, is found to have a more complex basis than the responsibility for research subjects.

Chapter 9 deals with fair credit in research and scientific publication, and Chapter 10 examines credit and intellectual property issues arising in engineering practice.

The epilog presents two stories of engineers who went beyond fulfilling their professional responsibilities to incorporating their values and aspirations into their work as engineers.

ORDER OF TOPICS AND USE IN COURSES

The interested engineer, scientist, or scholar may wish to begin by reading the entire Introduction or by simply skipping it. A detailed table of contents is provided as an aid for the general reader who wishes to read selectively, although each chapter does build on previous chapters.

If this book is to be used as a course text, the sections of the Introduction are best considered in concert with the early chapters. For example, Part 3 of the Introduction, on moral character and moral responsibility, is well considered in conjunction with the substantive discussion of professional responsibility and the engineer's responsibility for safety in Chapter 2 or 3. (A scheme for using the book in a single course is provided in the syllabus for *Real World Ethics*, one of the courses in engineering ethics available through the WWW Ethics Center for Engineering and Science (http://ethics.cwru.edu)). Cases and materials marked with "www" may be found in the WWW Ethics Center. The book does not presuppose any particular prior course of study, and its early chapters are accessible to all undergraduates.

Because the book provides a coherent guide to many topics within engineering and research ethics, it is suited to unifying the educational experience of engineering and science students who are learning engineering ethics by the "pervasive method," that is, having topics in engineering ethics and research ethics included in their science and engineering courses. Used for pervasive ethics education, Chapters 2 through 4, together with related case materials on the world wide web (WWW) and available on videotape, are suitable for use with first- and second-year students. The remainder of the book is best used in upper-level undergraduate and graduate courses.

When the book is to be used as a primary text in a freestanding course in engineering ethics or research ethics, it should be a course for upper-level or graduate students. Students will best understand the issues if they have some experience handling complex responsibilities. Many students enter college with such experience, but not all do. Summer work experience often provides a very useful experience on which to draw in class discussion.

The only topic I regularly address in my own undergraduate course that I have omitted from the book is the topic of academic honesty. I have omitted it because of my commitment to active learning and the realization that the approaches to the most effective approach to active learning about academic honesty requires linking problems and cases to specific policies and issues on one's campus. For example, where there is an honor code, it will be important to examine how that functions. If there is a student court, then it may be appropriate to spend some time on questions of procedural justice. Academic honesty is one of the first topics to cover in the pervasive method of teaching professional ethics. I find that the subject of research ethics provides a useful reprise for upper-class undergraduates and graduate students on such topics as plagiarism.

An appendix to this book discusses several major trends in philosophical ethics since 1980. To spare student readers the added expense of a larger book with an appendix that few of them would actually read, I have placed the appendix on the WWW. Engineers, scientists, philosophers, and social scientists who are interested in an explicit discussion of the philosophical position underlying this book will find it there. Here

I will simply say that active learning in professional ethics should in-volve students in hands-on/minds-on learning. Students should learn how to reflectively consider moral problems and moral standards and examine such standards with others of diverse backgrounds. Philosoph-ical work on topics such as trust, responsibility, and harm is useful in such reflection, but theories about how one might found ethics on rea-son alone are best reserved for courses in the history of philosophical ethics. [In practice, what often happens when terms such as "utilitar-ian" or "rights theory" is introduced in courses in professional ethics is that students get the ludicrous impression that they are expected to choose between considering consequences and rights (or duties or con-siderations of virtue) in making ethical assessments.] The view that the reflection that differentiates ethics from mere custom is *social* reflection, and that it is carried out with respect to particular problems and issues, rather than being the reflection of a lone scholar who considers action in the abstract, finds support in the challenges that many of the most distinguished philosophers recently have offered to the abstract and de-tached model of philosophical reflection. Annette Baier summarizes some of those challenges in the following terms:

> Bernard Williams and Thomas Nagel have both in their recent books[1] raised the question of what philosophical reflection [that is, detached, abstract consideration], especially that which Hume called "a distant view or reflexion" (T 538), does to what Williams calls our "confidence" in ourselves and our mores, and our personal projects and commitments. Is what Nagel calls "objective engagement" a real possibility for us, or will the attempt to be detached and reflective have the effect of detaching us from all engagements, destroying our confidence in any project, making all our concerns seem "absurd"? Will the philosophically examined life be found to remain worth living? Williams says "the ideal of transparency and the demand that our ethical practice should be able to stand up to reflection do not demand total explicitness, or a reflection that aims to lay everything bare at once...I must deliberate from what I am. Truthfulness requires trust in that as well, and not the obsessional and doomed drive to eliminate it" (p. 200). Though I welcome Williams's emphasis on the importance and fragility of confidence, and his reminder of the close link between the trusty and the true, I would amend his statement to "we must deliberate from where we are"; for, as he himself emphasizes, confidence and trust are social achievements. We may be able more successfully to combine self-trust with explicitness and reflectiveness if we can abandon the "forelorn solitude" of that singular philosophical thought which turns

each of us into "a strange uncouth monster" (T 264) and incorporate into
our philosophical reflections on morality more of the social and motiva-
tional resources of morality itself. For our form of life to be able to "bear
its own survey" (T 620), maybe both the life and the method of surveying
will have to change.[2]

ACKNOWLEDGMENTS

Many people have contributed to my understanding of engineering
ethics and research ethics and to the writing of this book. First are
the faculty with whom I have taught, especially Stephanie Bird, Larry
Bucciarelli, Peter Elias, Woodie Flowers, Nelson Kiang, Albert Meyer,
Igor Paul, Steve Senturia, Tom Sheridan, Leon Trilling, and guest lec-
turers in several courses: Stephen Chen, Randall Davis, Stephen Fair-
fax, Yolanda Harris, J. J. Jackson, Vera Kistiakowsky, Freada Klein, Mark
Kramer, Elizabeth Krodel, Judith Lachman, Jenny Lee McFarland,
Richard Petrasso, Steve Robbins, Andrew Rowan, Mary Rowe, Susan
Santos, Gerald Schneider, David G. Wilson, and the students in my
courses, some of whose scenarios appear in this book and whose projects
are available on the WWW.

I also thank the many people who aided the effort to teach engineer-
ing ethics at MIT and in the larger community, Mildred Dresselhaus,
Hermann Haus, Jack Kerrebrock, Paul Penfield, Robert W. Mann, Sheila
Widnall, and David Wormley.

Thanks to Kathryn Addelson, Pamela Banks, David Neelon, Stanley
Hauerwas, Jennifer Marshall, Helen Nissenbaum, Aarne Vesilind, and
David Gordon Wilson, who criticized parts of the manuscript, to Djuna
Copley-Woods who drew the diagrams, and to the students in my Real
World Ethics class in the spring terms of 1995, 1996, and 1997 who gave
responses to it as a text.

I especially thank my wise and tender husband, David Neelon, for the
sure pleasure of his company throughout this work.

NOTES

1. Bernard Williams, 1985, *Ethics and the Limits of Philosophy*. Cambridge, MA:
   Harvard University Press; Thomas Nagel, 1986. *The View from Nowhere.* New
   York: Oxford University Press.
2. Baier, Annette, 1986, "Extending the Limits of Moral Theory." *The Journal of
   Philosophy.* 77:538–545.

FOREWORD

I want to die proud of having been an engineer. Since that can happen only if we engineers behave ethically, and since I see a connection between this book and gracious professionalism, I am very enthusiastic about Dr. Whitbeck's effort to help us think effectively and somewhat pragmatically about professional ethics. Everyone, professionals in particular, must expect ethically complex situations to arise. When that happens, each of us badly needs a self-image that includes conviction that our intellect and heart can help make choices that will dramatically affect the course of events. That point of view will not materialize out of the ether. It must be nurtured and encouraged. This book will help seasoned professionals clarify their approach to their own behaviors, and this book can profoundly affect those who face a messy situation for the first time.

Caroline's arguments penetrate some of the fog around ethics. Most people think of it as an obscure topic belonging to an elite few who can spend their lives in deep and abstract thought. Even many professors of engineering regard ethics as a somewhat untouchable topic. "Students will never listen! Why waste our time and theirs?" Several have argued that post–high school is too late to influence students' proclivity to behave in society's best interest. I strongly disagree. Since I have spent most of my teaching career encouraging students to trust their own creative abilities, I have developed a thick skin about comments like, "You cannot teach creativity!" I do not debate that assertion. I think I know that one can unleash creative behavior by ensuring that it is overtly rewarded and by providing people with an assortment of "tools" which facilitate creativity. Likewise, after ten years of knowing Dr. Whitbeck and listening to her discussions, I am convinced that one can develop a self-image that includes self-confidence in dealing with ethically complex situations. I think that self-image is part of the foundation for a role as one of the protectors of society. It is essential to one who derives satisfaction from doing the thing that is right rather than easy or lucrative in the short term.

If students are told about an ethically complex situation and asked what course they would take if they found themselves in such a plight, they are quite likely to argue that they should call a press conference and blow the whistle on the bastards! Only after some discussion do they start to visualize the many scenarios that might accompany the choices made by the players. In a successfully guided discussion, they see that their creative and problem-solving talents are important resources and start to propose actions that minimize damage without "selling out." They start to synthesize solutions rather than judge the players. Thus, as Caroline argues, there is a strong parallel between the process of design and the process that should be used to guide one through fate's hammerlocks. The "problem" is ill-defined and resplendent with ambiguity and untruth; creativity and the wisdom to recognize what is important are critical, the iterative process of synthesis and analysis applies, and the solutions are not likely to be perfect, especially as judged by many stakeholders. One is not born comfortable with such a fuzzy and emotional process. Like design, one best learns it through a supervised opportunity to practice. This book provides such a guided opportunity.

The case studies provide very rich examples of successful and unsuccessful attempts to deal with ethical complexity. They illustrate that the right path is sometimes frightening and very rough. In the case of Roger Boisjoly and the Challenger disaster, he was forced to endure personal and professional persecution before being recognized as a most exemplary advocate of the "right thing" in an industry obsessed with the "right stuff." Mental experiments, classroom exercises, and personal introspection founded on Mr. Boisjoly's incredible story can be very productive. To borrow from the late Senator Everett Dirkson, an epiphany here, an epiphany there, and before long, we are talking real understanding. Interaction with Caroline has helped me understand what I think about when forced to confront ethicaly complexity. Thankfully, my ethics and religion are very simple and grow from the notion that we should all behave in a way that enhances the community good. I struggle with deciding the proper scope of "community." To me it includes animals, but what about plants? My most robust observation about "good" is that it is only a function of time until reelection, or it depends only on the time period over which the evaluation is performed. But given those vagaries, I find that Dr. Whitbeck has given me a nice road map for thinking about my actions. I recommend that you enjoy this book and allow it to make your brain hurt a bit to ensure that the message sticks. Many times, we can do well while doing good.

Woodie Flowers
Papalardo Professor of
Mechanical Engineering,
MIT

INTRODUCTION TO ETHICAL CONCEPTS

A clear understanding of the terms, concepts, and distinctions that people use to describe ethical problems and concerns helps in identifying what is ethically significant (or "morally relevant") in a situation. Understanding the ethical significance of the problems we face is the first step in responding well to them. Clear concepts and distinctions are also needed in the reflective examination of the ethical soundness of practices and customs. Standing up to such examination is what distinguishes ethical convictions from mere habits of thought.

The tendency to avoid ethical language is so widespread that even common terms for describing ethical situations seem strangely unfamiliar. Although avoiding ethical language may, in some circumstances, serve to reduce the defensiveness of those whose actions or policies are being questioned, it inhibits the understanding of ethical problems that commonly occur. The precise use of concept is essential for careful reasoning and clear communication in any field, and a consistent use of terms is required for parties to be able to recognize when they are agreeing, disagreeing, or addressing different subjects.

This introduction is intended both to clarify ethical terms and distinctions and to provide a general framework for considering ethical questions. If discussion of ethical terms is new to you, you may want to read through the main text, skipping the fine points that are set off to the side in smaller type. Along with the ethical terms discussed in this introduction, other, more specialized, ethical, legal, and technical notions are introduced as needed throughout the book.

The definitions of ethical terms in this introduction follow accepted English usage closely. Sometimes, when a word has several meanings, I use only one of these for the sake of clarity. I avoid stipulating new technical senses of words, however, for two reasons. First, the ethical distinctions marked in language express many important and subtle

1

distinctions that most readers will find they have been using all their lives without having thought explicitly about them. Second, part of my purpose is to prepare readers to discuss ethical problems, concerns, and questions with others who have never read this book, a goal that would be undermined by introducing jargon. Therefore, I focus on ethical distinctions that are directly applicable to addressing ethical problems in engineering and science.

The ethical distinctions discussed here are those marked in English. These distinctions are not precisely the same as those marked in other languages. These distinctions necessarily embody a cultural perspective, although I have tried to show some ways of expressing a variety of cultural and religious views on ethical matters.

A variety of distinctions have been drawn between the terms "moral" and "ethical." For example, philosophers often reserve the term "ethics" for the *study* of morality. Others, including many engineers, take "moral" to apply to private as contrasted with professional life. I make no distinction between "moral" and "ethical" in this book, precisely because of the multiplicity of distinctions in use; to use one of the distinctions would invite confusion with a host of others.

The discussion here is intended to provide a vocabulary rich enough to think about and address ethical problems and make ethical judgments. It is not intended to establish whether some act, motive, or character trait is ethically acceptable. I have chosen illustrations of ethical concepts I think are relatively noncontroversial. If you disagree – for example, if you think one of my examples of a human right is not a human right at all – understand that such a question is not supposed to be settled by this discussion. The examples are simply intended to make the concepts easier to grasp.

The problems addressed in this book arise primarily in engineering as it is practiced in the United States, Canada, Australia, and other technologically developed democracies. The point, however, is to understand ethical notions, whether or not English or some other language has ready terms for them. Ethical terminology changes over time. For example, although the term "right" was coined only in the modern period in relatively individualistic cultures, the notion of moral rights, and more specifically, of human rights, now finds widespread international acceptance.

To the extent that a notion is applicable only in a particular cultural or societal setting, it is important to recognize the cultural or political assumptions built into the notion. "Privacy" is sometimes claimed to be a notion that is applicable only in relatively individualistic societies; some languages, like Japanese, have no term for it. Even if only relatively individualistic societies emphasize privacy of the individual, what many Americans object to as violations of individual privacy may be objected to as unwarranted invasions of family or group life in other cultural settings. Therefore, discussions of subjects like the influence of technology on privacy may have some relevance for societies that do not emphasize the privacy of individuals.

The same general conditions of engineering and scientific practice hold for many technologically developed democracies. However, some specific conditions of practice vary. For example, although some states in the United States are now moving to require engineers practicing within their borders to become licensed, engineers employed in industry are exempt from such a requirement, and so the majority of employee engineers are not licensed in this country. In Canada all engineers must become licensed, and it is the engineering society for each province that has the legal authority to revoke licenses. In Australia there is at present no general requirement of licensure for engineers, although engineers must fulfill special requirements to be able to certify drawings. Australia is moving toward licensure, but since government is more centralized in Australia than in the United States or Canada, licensure will be administered rather differently from either the United States or Canada.

PART 1. VALUES AND VALUE JUDGMENTS

VALUES DISTINGUISHED FROM PREFERENCES

The question of what is good or bad, better or worse, desirable or undesirable is a question of merit or worth. It is a question of **values** and it calls for a **value judgment**. A value judgment is any judgment that can be expressed in the form "X is good, meritorious, worthy, desirable" or "X is bad, without merit, worthless, undesirable."

The first point to consider is the difference between being desir*able* or worthy in some respect and simply being desired, liked, or preferred by some person or group. This distinction is crucial to the later discussion of ethical judgments and standards for engineering practice. Consider these statements:

"I like fried peppers."
"I am unalterably opposed to having cats in the neighborhood."

Some expressions of dislike are not statements at all. For example, "Cough syrup, yuck" is not a statement. A statement has truth value; it can be true or false. However, if someone said "Cough syrup, yuck," it would be reasonable to surmise that the speaker dislikes taking cough syrup.

These are **statements of preference**, that is, of liking and disliking. Statements of preference are not judgments about whether something is good or bad in some respect.

Unlike a value judgment, such as, *fried peppers make a good side dish*, the statement of a preference, such as "I like fried peppers," is an assertion about *the speaker* rather than about fried peppers. More specifically, it is a statement about the speaker's feelings, views, or attitudes toward the thing in question. Statements of preference are false only if they misrepresent *the speaker's feelings*.

It is normal to feel repugnance at wrongdoing, but the strength of one's feelings are not a reliable guide to the gravity of an offense. As people mature they learn to distinguish between their feelings on a subject and their moral judgments. For example, someone may believe that, ethically speaking, shooting a person is much worse than shooting a dog. However, if his own dog had recently been shot, but he had never personally known a person who was murdered, he might well experience stronger revulsion when hearing about the shooting of a pet than when hearing about the murder of a person. Like this speaker, a person may know the origins of his preferences and attitudes and may give *causal explanations* in terms of psychological factors that have contributed to their development. For example:

"We always served fried peppers at celebrations when I was growing up."
"When I was a young child, my closest friend was attacked by a cat."

Alternatively, one may analyze her preferences to identify more precisely what it is she likes or dislikes:

"I can't stand the sound of cats fighting."

Such a person may even give you reasons for thinking that what he prefers is desirable or at least desirable for him, such as:

"Cats carry disease."
"I am extremely allergic to cats."

However, the speaker need not give any reasons for a preference. For some matters, such as preferring one flavor of ice cream to another, people usually do <u>not</u> have reasons for their preference. When you state your preference, you are stating your attitudes or feelings, not giving a reasoned judgment. A person may have a strong preference for something without believing it fulfills some high standards or brings about some good. He may not even know how he came to prefer what he does.

If something is claimed to be *good* or desi*rable*, one makes a statement about *the thing* that is claimed to be good, rather than about *the person* who likes it. As Aristotle first observed, to say that something is good or desirable is to say that it has qualities it is *rational* to want (in a thing of that sort). For example, a good knife is one with the properties it is rational to want in a tool with one blade used for cutting, such as being sharp, well-balanced, and comfortable to grip. To claim that something has the qualities that it is *rational* to want in that sort of thing is to claim that there are good reasons for wanting it.

Given the differences between value judgments and statements of preferences, you may expect that others expect you to back up your judgments and preferences in different ways. If you make a value judgment, others are likely to ask you for the reasons you judge it rational to want (or reject) whatever is the object of your judgment. If, on the other hand, you merely state your preference, you need give no further reasons for your liking or disliking. You may or may not *have* reasons underlying your preference. Value judgments on very major questions and the commitments that usually attend such judgments – such as "this is a good (or "the right") person for me to marry" – are not likely to rest on a few simply stated criteria as is the judgment that something is a good chair or knife. It may take a great deal of time and thought to make explicit the criteria for a major value judgment.

As children, people develop habits, ways of thinking and acting, that reflect the value judgments of the adults who raised them, the culture in which they were raised, and their own particular life experiences. Part of the work of adolescence is to begin the examination of those habits and see which are justified, morally or otherwise. Universities and high-tech workplaces are environments in which individuals typically encounter those with habits and values very different from their own. Examination of one's own habitual ways of thinking and acting may be stimulated by the contrast with the habits of others. It requires maturity, however, to simultaneously show a day-to-day tolerance for others with very different habits *and* to take seriously that one's own and other's actions and values must be critically examined.

Some thinkers argue against the use of the term "values" because it may suggest that all values are somehow on a par. Worse yet, it may suggest that all values are reducible to monetary value or a measure of "willingness to pay" for something. Such a notion of a value is virtually indistinguishable from a preference and is, hence, purely subjective. One will find many careful thinkers who avoid the term "values" and speak instead of "goods" or "types of flourishing" when discussing desirable outcomes. The term "values" is so widely used, however, that it would only invite confusion to try to avoid using it. If the term "value" is used carefully as a noun or an adjective, the distinction between values and preferences will not be blurred.

The use of the term "value" as a verb, "to value," is confusing because it is virtually synonymous with either "to like" or "to assign a monetary value to." Using "value" as a verb obscures the distinction between value judgments and mere statements of preference. I never use "value" as a verb. If a verb is needed, I use "evaluate," meaning assess according to standards. This preserves the connection of value with reasons or standards and distinguishes it from mere preference.

TYPES OF VALUE AND VALUE JUDGMENTS

There are different types of value and value judgments. Both works of art and naturally occurring objects and events may be judged in terms of **aesthetic value.** Words like "beautiful," "harmonious," "elegant" and "engaging" are terms of aesthetic praise. Words like "ugly," "banal," "dull," and "lopsided" are terms of aesthetic scorn.

Statements along with hypotheses, research studies, theories, and designs for experiments are also judged to be good or bad in terms of what are sometimes called **knowledge values** or **epistemic values**. These include truth, informativeness, precision, accuracy, and significance. Research is judged not only by whether it reveals a relationship that is unlikely to have occurred by chance – is "statistically significant" – but by the importance of its implications. Research is also judged in terms of its fruitfulness, that is, of the further lines of inquiry the research questions or results suggest. Hypotheses are judged in terms of plausibility, the scope of the phenomena explained, and testability.

Plans and strategies are common objects of **prudential judgment**. When people speak of a good (prudent or effective) strategy or a bad (stupid, short-sighted) plan, they are making a prudential judgment about the efficacy of the plan or strategy in question, that is, whether it will achieve certain ends. Behind most prudential judgments are other value judgments that certain ends are *worth* achieving.

A special case of an end that is generally assumed to be valuable is survival, either biological survival or continuance as a member of some group. A plan or idea is generally judged imprudent or stupid not only if it is ineffective, but to the extent that it neglects the survival or well-being of the agents. People do speak colloquially of "survival value." When two people disagree in their prudential judgments, they may be disagreeing about the dangers in some course of action or the importance of what is endangered. For example, consider the warning

That which is valuable as a means rather than "as an end," that is, for its own sake is said to have instrumental value. That which is valuable in itself (rather than merely as a means) is called an intrinsic value. (The same thing can be valuable as both a means and an end, of course; for example, health is both good in and of itself, but it also enables a person to achieve many other goods.) Prudential value, unlike aesthetic value, epistemic value, or the other types of value discussed in this section, is instrumental rather than intrinsic since prudential value is the usefulness of the thing in question in achieving some distinct end.

"If you want to survive in this organization, you will not report corruption. It would be stupid to stick your neck out." If one believed the speaker, one might decide to leave so corrupt an organization.

The last major type of value I will discuss before turning to ethical value is **religious value**. The terms of evaluation include "sacred" and "holy" as contrasted with "profane" and "mundane." Purely religious standards are often applied to people, writings, objects, times, places, liturgies, rituals, stories, doctrines, and practices. Religions that emphasize the importance of doctrine are called "doctrinal"; others emphasize liturgy, the order of worship. Some religions understand life in terms of sacred stories. Other religions emphasize nonliturgical practices, such as forms of yoga or meditation. Some emphasize care of less fortunate people or compassion toward all sentient beings (beings that can feel pain). One emphasis may coexist with others. (These differences are noted because a surprising number of philosophers write as though doctrine were the central concern in religion, but for many religions, it is not.) Some emphases change over time. For example, in Judaism before the exile, a place – the Temple at Jerusalem – had central importance. After the exile, when people could no longer go to the Temple, scripture – the Torah – became central.

Most existing religions, and all major world religions, uphold ethical as well as religious standards. These ethical standards apply to moral agents – to their character traits, motives, or actions. Religions vary somewhat in their relative emphasis on spiritual and moral virtues of individuals, an ideal of family life, and on the faith or practice of the people as a whole. Confucianism puts great emphasis on the family, for example, and largely defines the virtues of individuals in terms of their place in the family. Family and caste identity have had a defining role within Hinduism. Buddhism, in contrast, emphasizes enlightenment of the individual. Judaism emphasizes the relation of the whole people of Israel to God, and praiseworthy individuals are those who strengthen the relation between God and the people of Israel. Because Christianity emphasizes individual salvation, it is generally regarded as more individualistic than Judaism, notwithstanding a continuing emphasis on the community of the faithful or "the Church." Islam emphasizes the duty to form an equitable society where the poor and vulnerable are treated decently.[1]

Notwithstanding differences concerning the primary social and spiritual unit, many religions share ethical norms and even some underlying convictions that support the practice of those norms. For example, Hinduism and Buddhism hold that karma, the total effect of a person's actions in successive phases of existence, determines that person's destiny. All major world religions have some version of the admonition to treat others as you would want to be treated: the "Golden Rule."[2www]

In addition to generally applicable ethical norms, religions often offer spiritual guidance to their members about what they as individuals are particularly called to do. The native American practice of embarking on a "vision quest" to discern one's life path is an example of a means of seeking spiritual guidance. The Middle English word from which the term "vocation" derives means a divine calling, although questions of vocation may be addressed by exclusively secular means, such as aptitude tests.

One type of value may be relevant to another. For example, aesthetic criteria, such as beauty and symmetry, are commonly held to enter the assessment of scientific theories. Conversely, many argue that great art gives a profound insight into reality, which brings aesthetic value close to religious, or perhaps scientific, value. So, although I have distinguished various types of value here, it is an open question whether there are fundamental connections among them.

Notice that all of the types of value discussed differ from **market value**. When one assesses market value, one is <u>not</u> making a value judgment of what is good or bad in some re-spect. Rather, one is simply deter-mining the price at which the sup-ply of an item equals demand for it. For example, we all need breathable air for health and survival – health is commonly regarded as a funda-mental good. Because there is no scarcity of air of breathable quality in most areas, no one needs to buy it. Therefore, breathable air has no market value.

Just as valuable items, like clean air, may have no market value, so high "market value" may attach to items that are not good by any rea-sonable standards. Market value de-pends on the relation of supply and demand. Thus it may depend on the strength of preference of those who have the means to pay for an item and the willingness of those who have it, or can make it, to sell it. An addictive and physiologically destructive drug with analgesic or euphoric properties might have high market value. Such a drug would not be "good" even in the sense of having properties it would be rational to want in an analgesic or euphoric drug.

> The term "economic value" is sometimes used as a synonym for "market value," but not always. "Economic value" often means the usefulness of the object in question for creating prosperity, and thus it is a type of instrumental value. The economic value to a country of having a system of transportation and sanitation is not the price of these systems if sold, but rather the prosperity the systems help create.
>
> Notions like "nutritional value," "sanitary implications," "security implications," or even "entertainment value" are also types of instrumental value. Nutrition, health care, and security are the sorts of things that humans have an interest in being able to obtain.

ETHICS AND ETHICAL RELATIVISM

Having examined other sorts of value and value judgments, we will now consider ethical values and ethical evaluation. As Amelie Rorty observes, it is not always a simple matter to classify a value judgment as being aes-thetic, moral, prudential, etc. Furthermore, the distinction between types of values has been drawn only in modern times. Aristotle, for ex-ample, does not distinguish between types of value.[3] When one does distinguish between types of value, consideration of one type often has implications for others. Prudence in fulfilling one set of moral respon-sibilities is necessary if one is to be able later to fulfill others. Prudential considerations along with ethical ones arise in many problems and sit-uations discussed in this book. Nonetheless, it is useful to attend to the characteristics of value judgments that are identifiably ethical judgments.

One of the first questions that many people raise about ethical values and judgments is of how generally they apply: whether ethics is "relative" so that ethical judgments have limited applicability.

Some authors use the term "ethics" or "morality" for any code of behavior, even one that does not claim to have moral justification. For example, Robert Jackall in *Moral Mazes* describes what he calls a corporation's "ethics" or "morality" and takes it to include such judgments as "What is right is what the guy above you wants from you"(p. 6). Such a judgment is about the best (i.e., most effective) way to survive in the organization but does not pretend to be a statement about what is morally/ethically justified. Examination of such codes of behavior may be important since they affect the opportunities for moral action, but they have no ethical import. I follow the usual practice of calling a code of behavior an ethical code only if it is supported with ethical justification.

"Ethical relativism" or simply "relativism" is used for several distinct views. First it is used for a view, more precisely called "ethical subjectivism." Ethical subjectivism holds that whether it is right or wrong for a person to do a certain thing in a given situation is determined by whether that person thinks the act is right or wrong. This view represents ethics as lacking in objective standards. It is difficult to have much of a discussion about the reasons for thinking some action is right or wrong if a person's moral beliefs determine whether it is right or wrong for that person, and the beliefs themselves cannot be subjected to any standards.

It is important to distinguish between ethical subjectivism and another view also called "relativism" or "cultural relativism." Cultural relativism with regard to ethics is the view that ethical judgments, rules, and norms reflect the cultural context from which they are derived and cannot be immediately applied to other cultural contexts. Cultural relativists put the burden of proof on those who think that they can generalize from one social context to another.

Many, like philosophers Annette Baier and Alasdair MacIntyre, who do not consider themselves relativists, argue that moralities are cultural products constructed by particular people in particular historical contexts and must be understood in relation to those societal contexts. For example, the Hippocratic oath specifies extensive duties of physicians toward those who taught them medicine. In this oath, physicians pledge to respect and care for their teachers as for their parents. The societal context in which these duties of physicians were formulated was very different from that in industrialized nations today. It is implausible that physicians would have the same duties toward their teachers as did the physicians in ancient Greece. This observation does not mean that there was no moral validity to the duty when the oath was first formulated. What makes the difference is not a person's opinion, but the social reality in which the person participates. This social reality provides the context in which people subject their own loyalties, convictions, and self-understanding to examination and criticism, examination and criticism that transforms their habits of thought into genuine moral convictions.

Cultural relativism applied to a single culture over time yields a modest relativism, according to which it is simplistic to judge an action in some other period solely by today's standards. This does not mean that an action can be criticized only by the criteria commonly used in that period, however. For example, informed consent for medical experiments is a standard that has developed in industrialized democracies only since the Second World War. The implicit prior standard was, "First do it [the experiment] on yourself." Someone who used the prior standard conscientiously in 1940 is not subject to the same moral criticism as would a person today who knows or should know about the informed consent standard. Nonetheless, the informed consent standard is arguably a superior standard; we would think highly of someone for seeking his subject's informed consent for experiments in 1940.

OUGHT, GOOD–BAD, RIGHT–WRONG

The term "ought" is sometimes used to mean what one should do, *other things being equal.* For example, "You ought to avoid bad company" means that, other things being equal, you should avoid bad company, not that there are *no* circumstances in which you ought, ethically speaking, to be in bad company. Sometimes "ought" means what one should do, *all things considered* – as in "In these circumstances what you ought to do is start over" or "Given the situation, I think you ought not press your right to x." Often, as in the above examples, when the general case is being discussed, "ought" without further specification is understood as "necessary, other things being equal." When a specific instance is being discussed, "ought" is understood as "ought, all things considered." That is how I will use the terms here: When I discuss a specific case, "ought" without further qualification means "ought, all things considered." When I discuss a general type of situation, "ought" without further qualification means "ought, other things being equal."

Ethical evaluation of an act or a course of action can take the form of a judgment about whether (or the extent to which) the action was a good or a bad thing to do. Since **ethical justification** is what makes a code of behavior an *ethical code*, and justification turns on the moral evaluation of acts, motives, and agents, ethical evaluation is a central topic for ethics. (Notice that one can also inquire into the justification of any moral evaluation. One can always ask "Why is it better to do X than Y in these circumstances?" Therefore, the relationship of evaluation and justification is reciprocal.)

Ethical terms, the terms we are examining in this introduction, provide that language for ethical evaluation and justification. For example, "signing the peace accord was a *good (compassionate, responsible, beneficial)* thing to do" makes reference to the intended consequences and the virtues displayed in the act in question. Ethical evaluation of an act can also be in terms of the rightness (or wrongness) of the act, that is, whether (or the extent to which) it is "the right thing to do." The two types of ethical judgment, good/bad and right/wrong, are similar in some respects. The presence of "the" in "the right thing to do" rather than "a" as in "a good thing to do" suggests that "the right thing to do" is the *only* morally acceptable response. However, speakers do not always mean that there is only one acceptable response. Speaking of an act as right or wrong usually signals that the speaker's justification for her ethical judgment rests on an appeal to moral rules, rights, or obligations. Speaking of good and bad things to do usually signals a consideration of the total consequences or the display of virtues, such as kindness or wisdom.

A NOTE ON NORMATIVE ETHICAL THEORY

In considering an ethically significant problem one should consider *all* of the morally relevant factors; this introduction discusses concepts for expressing the full range of morally relevant considerations. When normative ethical theories are brought into discussions of practical and professional ethics, they are commonly misrepresented as directives to *restrict* the morally relevant factors one considers. A **normative ethical theory** is a philosophical theory about what constitutes the ultimate or most basic ethical norms or standards. Some of the best known of these offer a basic criterion that is supposed to be something to which all rational beings would agree. Two of the best known are "one should act so as to achieve the greatest good for the greatest number of individuals" – John Stuart Mill's utility principle – or "act so as to make your action a general rule for action" – one formulation of Immanuel Kant's categorical imperative. [4] Accepting a particular normative ethical theory is often misunderstood as implying that in evaluating an action, one considers only the sort of factor that one considers most basic. For example, readers of works on professional ethics frequently come to believe that if one is a utilitarian, one considers only consequences of acts, and if one is a Kantian, one considers only whether some act conforms to a moral rule. This representation makes normative theories appear to be techniques for ethical evaluation. That they are not is demonstrated by the existence of positions such as "rule utilitarianism." Rule utilitarianism holds that the greatest benefit is achieved by people following moral rules. (For example, people should keep their promises because if people broke their promises whenever they thought it would be generally beneficial to do so, the trust needed for cooperative ventures would break down creating a generally worse situation.) Because normative ethical theories are about the *nature of ethics*, not techniques for moral evaluation, such theories do not require explicit discussion here. Furthermore, since 1980 a growing number of distinguished philosophers, including Kathryn Addelson, Annette Baier, Stanley Hauerwas, John Ladd, Alasdair MacIntyre, John McDermott, Amelie Rorty, Richard Rorty, Bernard Williams, and Iris Young, have argued that normative theories that attempt to found ethics on reason alone are attempting the impossible. Some of these are also those who argue that moral reflection and deliberation should be recognized as taking place within an historical context and that philosophers ought to aid the process of reflection but not attempt to give a grand theory of what ethics is.

In writing this book I make use of the work of philosophers like Kant and Mill, just as I draw on more recent philosophical work on responsibility, harm, and trust, but I do not discuss alternative normative theories themselves.

MORAL AND AMORAL AGENTS

Acts, agents, and the character or motives of agents are the objects of moral evaluation. However, only certain agents have their acts, character, or motives morally evaluated. For example, the statement "the storm was responsible for three deaths and heavy property damage" means that the storm caused these outcomes. Although the storm was the agent of destruction, the actions of the storm are not subject to *moral* evaluation. The storm is not guilty of murder or even manslaughter. Those whose actions, character, and motives can be morally evaluated are called **moral agents**. A competent and reasonably mature human being is the most familiar example of a moral agent. In contrast, most "lower" (that is, nonhuman) animals are generally understood to be **amoral**. Saying they are amoral is to say that morality is not a factor in their behavior, and, therefore, questions of morality are not appropriate in evaluating them and their acts.

To say that lower animals are not capable of acting morally or immorally is not to deny that there are moral constraints on the way moral agents should treat them. Moral constraints on the way lower animals are treated is a matter of the animals' **moral standing**, that is, their intrinsic moral worth, not their **moral agency**. I discuss moral standing in Part 2, in the section Moral Obligations, Moral Rules, and Moral Standing. Any moral agent has moral standing, but the prevalent view is that some beings are not moral agents yet have moral standing. For example, it is generally held that it is wrong to be cruel to animals – even though they are incapable of moral action.

Human beings and human groups such as nations are the most familiar **moral agents**. However, some other species, such as porpoises, are often alleged to qualify as moral agents, even though they are not human. Boulle's book *The Planet of the Apes* portrays apes as moral agents. Science fiction often describes nonhuman extraterrestrials as persons and moral agents. Various religious traditions speak of beings, such as angels or dakinis, who seem very much like people and act as moral agents. Humans may be the most common example of moral agents, but they are not the only possible example.

Moral agents are not necessarily morally good individuals. They are those who can and should take account of ethical considerations. Moral agents are those of whom one may sensibly say that they are either moral or immoral, ethical or unethical, in contrast to the amorality of most other beings.

Common terms of moral praise for agents include "good," "a person of high moral character," and "virtuous." Particular character traits that are praised as moral virtues are "kindness," "honesty," "courage," and "bravery." Acts are judged as right and wrong according to three criteria:

1. the nature of the acts – e.g., killing is wrong.

2. the specific circumstances of a particular act – e.g., Arthur's *unprovoked* assault on Cecil was wrong.
3. the motives with which the agent committed the act – e.g., Cedilla's criticism was destructive and motivated by hostility rather than a sincere attempt to improve performance.

CONSEQUENCES, HARMS, AND BENEFITS

The injury or benefit an action causes others is a morally relevant consideration. An action may directly or indirectly help or harm others. Someone is harmed directly, for example, by being run over. A person is harmed indirectly if something that she cares about or in which she has an interest is harmed or diminished. Because indirectly as well as directly injuring or benefiting a person is morally significant, values and interests of all sorts – not only moral, but also religious, aesthetic, epistemic, and prudential – are often relevant to moral evaluation.

The formal technique of **cost-benefit analysis** may be helpful in clarifying the trade-offs involved in exchanging one harm or benefit for another for a special class of problems for which the *consequences considered are ones that can be assigned arithmetic quantities.* (The same action may produce both harms and benefits, of course. For example, some currently used measures to control bacteria in the water supply introduce minute quantities of carcinogens into the water.) In cost-benefit analysis one compares different courses of action by multiplying the probability that a given course of action will produce some outcome by the magnitude of the harm (or benefit) of that outcome, and comparing this quantity to the quantity resulting from alternative actions.

In the past, product design decisions have been made by assigning a dollar amount to a human death for purposes of having a figure that could be compared to the cost of making a product safer. That proved to be short-sighted because it does not take account of actual damages in liability judgments and the costs of negative publicity. Aside from those prudential concerns, the assignment of a dollar amount to human life raises ethical questions, as we shall see later in the discussion of rights and human rights. It also points to a difficulty in quantifying harms. Today, the quantification problem is often gotten around by making a substitution of "willingness to pay" (to avoid the harm or receive the benefit) for judgments of degree of harm and benefit.

If the type of harm or benefit is held constant, the task is somewhat simpler. For example, one might compare business plans as to the increase in market share for a product that is expected to result from them. When the harm or benefit is held constant, the technique is called "risk-benefit" rather than cost-benefit analysis. (Assessments of the probabilities that a course of action will achieve a given result may be very difficult to make, however. The field of risk assessment has developed many sophisticated means for estimating these probabilities. Such assessments often raise subtle value questions because they focus attention on some consequences rather than others.)

The *probability* that a given course of action will produce some harm multiplied by *degree* of harm defines **risk, in the technical sense.** This notion of risk is a bit different from several other senses in which the term is used. "Risk" is commonly used to mean a **danger** or **hazard** that arises unpredictably, such as being struck by a car or capsizing in a boat. The "unpredictable" element in this colloquial sense of risk links it to the notion of an accident. The term "risk" is also used for the *likelihood* of a particular hazard or accident, as when someone says, "You can reduce your risk of capsizing by sailing only in light or moderate winds."

Risk analysis, risk assessment, and risk management use the technical sense of "risk," that is, the probability or likelihood of some resulting degree of harm. In technical risk analysis one focuses on the resulting harm and not just the harmful event. So one would speak of the *risk of death by drowning or exposure* as a result of capsizing, rather than simply of the *risk of capsizing.* The harms (or benefits) often considered are increased (or decreased) probability of death ("mortality risk") or monetary loss (or gain). Using the technical notion of risk, one can compare, say, the relative chance of dying when traveling between two points by automobile and by airplane. One can also compare the risks associated with harms of different magnitudes. For example, consider two monetary risks: the rather likely event of losing money in a broken vending machine and the rarer event of having one's money stolen in a holdup. In some locales there is a greater risk of monetary loss from a malfunctioning vending machines than from being held up and robbed.

The difficulty in finding a nonarbitrary way of quantifying some harms is important to bear in mind when using either cost-benefit or risk-benefit analysis. For example, in the case of monetary loss to vending machines as compared with monetary loss to robbers, we did not consider the greater emotional trauma associated with being held up. Assigning an arithmetic quantity to such trauma is very difficult, and so consideration of trauma is easily passed over in favor of considering monetary loss only.

(Recall the saying that if your only tool is a hammer, you will see every problem as a nail.) Employing cost-benefit analysis and the technical sense of risk should not lead to ignoring harms that cannot be quantified in favor of those that can. It would be a mistake to conclude that someone was behaving irrationally in taking greater precautions against being held up than against using malfunctioning vending machines, even if the risk of monetary loss from machines was greater.

Comparison of different sorts of harms or benefits is difficult, not only because in a pluralistic society there are different notions of the good life and what promotes or frustrates it, but because a specific harm will have different implications for people in different circumstances. Consider whether it is worse to increase one's chance of death by 25% or to be painfully disabled for ten years of one's life. When such choices are made in healthcare, we say that the individual patient has the right to make them. Similar decisions, such as decisions about what side effects to tolerate to purify the public water supply, must be made for the general population. A measure or estimate of the degree that each consequence would be *preferred* by most people is used. As mentioned earlier, those preferences are often quantified as the dollar amount that people would be willing to pay to achieve the benefit or avoid

In 1823, Jeremy Bentham proposed the utilitarian calculus in an attempt to provide a moral justification for legislative reform. This calculus, which was later refined by John Stuart Mill, requires the measurement of "utility," by which Bentham meant the property that tends to produce the benefit of pleasure or happiness. However, Bentham did not formulate a program for satisfactorily measuring utility.

In 1906 Vilfredo Pareto, in *Manuale d'Economia Politica*, showed that for classical economics all that was needed to measure utility was a statement of preferences by the individual or group whose interests were in question. This measure of preferences, however, provides no generalized measure of the amount of either pleasure or happiness for humanity, but only the preferences of the groups and individuals who are sampled.

the harm. However, as we have already seen, preferences are subjective and are strongly influenced not only by value commitments but by personal history. They are not a measure of magnitude of harm or benefit. Furthermore, measuring harms and benefits in terms of willingness to pay ignores the fact that people vary in the importance that money has for them and their *ability* to pay. Thus two people might deeply want an expensive medical treatment, but one could readily pay for it while, for the other, having the operation would mean sacrificing all the family assets.

Risk-benefit and cost-benefit calculations may obscure other morally significant factors, such as whether some measure harms one group while benefiting another. When those who are harmed are different from those who are benefited, the concept is called "**risk shifting**." Even if the net risk is lessened, there remains an ethically significant question of the fairness of the shift of risk. It is important to understand the limitation of a technique to use it responsibly.

PART 2. MORAL RIGHTS AND MORAL RULES

MORAL RIGHTS

Along with the concepts of benefit and harm, one of the most familiar concepts in contemporary ethical discussions is that of a **moral right**. A right is a justified claim, entitlement, or assertion of what a rights-holder is due. For a person to have the moral right to have, get, or do something, there must be a moral basis or justification for the claim. These bases or justifications are different for different categories of rights. We shall see that "human rights" is a name given to those rights that all people have because they are people. Rights possessed only by some are called "**special rights**." For example, if I have promised that I will drive our car pool in February, then you have a special moral right to be driven by me in February. Special rights may be acquired through agreements or contracts, or through (chosen or unchosen) relationships – for example, "parental rights." In this section we will examine four major contrasts in the categorization of rights: alienable versus inalienable rights, human rights versus special rights, negative rights or liberties versus positive rights, and absolute rights versus prima facie rights.

Moral rights, along with moral obligations and moral responsibilities, constrain how far a person may go in seeking to improve an outcome. For example, suppose you find yourself in some sort of emergency where you can act to save one person's life or to save four (other) people's lives. (Other things being equal) you ought to save the four people, rather than one. However, the greater value of four lives as compared with one would not allow you to take one life to save four others. Thus it would not be morally permissible to kill one person to harvest that person's organs and transplant them into four people who each need one of the organs to survive. In contrast, although both human life and great art have value, art does not have moral rights. Therefore, it would be justified to destroy one great painting to save four others – for example, by using the first to wrap the other four.

There are **legal rights** as well as moral rights. Although an effort is often made to bring the force of law behind some moral right by making it a legal right, moral rights must be distinguished from legal rights. There is no contradiction in saying that a person has a legal right to do something but not a moral right to do it or in claiming that some laws are unjust. Laws that treated enslaved people as property violated the moral rights of those who were slaves. The argument given to justify slavery in the United States was that the Constitution guaranteed rights only to citizens. The law did not recognize slaves to be citizens and so did not accord them **civil rights**, that is, the legal rights of citizens. Furthermore, the law regarded slaves as the property of others. The Fourteenth Amendment to the U.S. Constitution provided that former slaves are citizens and all citizens possess the right to life, liberty, and property (although only men had the right to vote) and that naturalized citizens have the same rights as native-born Americans. The same year, 1868, the Burlingame Treaty was signed. It denied the possibility of naturalized citizenship to Chinese-Americans, although it permitted free immigration between China and the United States. In 1882, Congress passed the Chinese Exclusion Act, the first federal law preventing immigration to the United States of a specific ethnic group, and it was not repealed until 1943. A Constitutional amendment to accord voting rights to women was passed in 1920. The Civil Rights Acts of 1964–65 legislated against all forms of discrimination based on race, sex, religion, and national origins.

The term "human rights" is one that Eleanor Roosevelt brought into widespread use. Previously these rights were called the "rights of man" (or sometimes, "**natural rights**"). She chose "human" as a more inclusive modifier. There is now international and cross-cultural agreement that all people have some rights simply because they are people. Notice that it is being a person rather than being biologically human that is crucial. A culture of human tissue would be both alive and human, but it is not a person and no one would claim it has human rights. The topic of personhood is one we will revisit later in this introduction.

The view that there are human rights gained wide acceptance in the eighteenth-century Enlightenment. It strongly influenced the U.S. Declaration of Independence, the framing of the Constitution, and the Bill of Rights. The term "right" itself dates only from the seventeenth century, however.

Arguments showing a basis for human rights are said to provide a "vindication" or "warrant for" human rights in general. One such argument is given by John Ladd. Ladd argues that human rights are the claims that need to be honored to provide a moral environment in which people are able to meet their moral responsibilities and maintain their moral integrity.[5]

The writers of the Declaration of Independence stated that human rights are a divine endowment. The view that human rights are divinely given may be unconvincing to those without religious convictions or suspicious to some who have religious convictions but doubt the religious authority of the writers of the Declaration of Independence.

Engineers and scientists confront issues of human rights directly when they face the requirement to obtain the informed consent of any person who is to be an experimental subject in their research.

The consideration of human rights provides a necessary backdrop for discussion of professional ethics. The international recognition of human rights provides endorsement of an ethical standard for ethical behavior that transcends cultural differences. Human rights are implicitly considered in formulating responses to a broad range of problems. Among those problems are the ethical problems of engineers that are the focus of this book.

The notions of a **moral rule**, and that of **virtue**, which will be discussed in the following sections, have been explicitly used in a larger range of cultures than has the notion of a right. Virtually every ethical and major religious tradition employs some counterpart of the notions of virtue and moral rule. Traditions vary on the content of moral rules, of course, and on the characterization of particular virtues, and on the relative importance of one moral virtue as compared with others.

Discussion of rights has a particular prominence in comparatively individualistic societies such as the United States, which in the late twentieth century is sometimes described as a "culture of rights" as contrasted with Japan, which is sometimes described as a "culture of duties."

In a pluralistic society with many different subcultures, people may agree more readily on what each person is due rather than on what each person owes others. Members of different cultures may agree that elderly people deserve or even have a right to certain care but differ on the moral duties of various family members, churches, communities, and the state in providing such care.

Philosopher Annette Baier points out that people make claims and give moral justifications for them in every human group with any social organization. Therefore there are claims with moral justification in every society. Since rights are justified claims, then there is an equivalent of the notion of a right in every society, even in those that do not have a ready term for justified claims. Cultures that see basic moral considerations in terms of responsibilities, virtues, obligations, and duties have the equivalent of moral rights, because the moral requirements they do recognize provide moral justification for certain claims of individuals.

Theologian Stanley Hauerwas argues that the language of rights does not do justice to the moral convictions of believers. For example, some religions prohibit murder because of the conviction that only God may legitimately take a human life, rather than because of a belief that the right to life is inherent in individuals. Nonetheless, Hauerwas subscribes to the formulation of human rights in Baier's culturally generalized sense.[6]

Many argue that there has been an excessive emphasis on moral rights in recent decades and neglect of other ethical notions, especially within the United States. Philosophers such as Annette Baier, John Ladd, Alasdair MacIntyre, and Martha Nussbaum; legal theorists Clare Dalton and Joel Handler; and psychologist Carol Gilligan have variously argued this overemphasis stems from

- the assumption that the moral evaluation of an act can be made without reference to the larger moral context in which the action occurs,
- a view of human relationships as being tenuous or adversarial,
- disregard of the fragility of human existence.

The objection to an excessive reliance on the concept of a moral right does not imply that the concept is useless, however.

The Declaration of Independence clearly rests on the assumption that human rights exist: All persons are created equal, for all are endowed with certain "inalienable rights." In the strongest sense, to say that a right is **inalienable** means that it cannot be taken away by others, traded away by the person, or forfeited as a result of the person's actions. In a weaker sense, it means that the right cannot be taken or traded away, but it could be forfeited through the person's actions. Although citizens have the right to vote, a convicted felon forfeits the right. However, any trading away of the right to vote is morally invalid. In the weakest sense, to say that a right is inalienable means only that others are not justified in removing or abrogating that right. Thomas Jefferson and the framers of the Declaration of Independence regarded such criminal punishments as imprisonment and executions just, even though these involve the forfeiture of liberty or life.

Today, the inalienability of rights mentioned in the Declaration of Independence is interpreted as meaning one cannot trade them away. The inalienable right to liberty is generally agreed to mean that a person cannot make a morally valid agreement to sell himself into slavery. In view of the widespread seventeenth-century practice of making agreements to be an "indentured servant," it is not clear that the framers of the Declaration thought inalienable rights could not be traded away, at least temporarily. In this book the term "inalienable" will be used to describe a right that others cannot take away and that one cannot trade but that *can* be forfeited through one's own actions.

Rights that may be removed are called **alienable**. You may give up your ownership of a car by selling it, for example. This possibility illustrates that a property right, unlike the right to liberty, is alienable. Furthermore, the general right to property, which forbids that one's property be taken without compensation, may be lost if the property is obtained as a result of illegal activity.

Rights, even inalienable ones, need not be exercised. Rights are *justified claims*. Those claims may or may not be pressed. Rights can be either **exercised** or **waived**. One may fail to exercise a right for many reasons, including not getting around to it. For example, if you obtain a driver's license (a special legal right to drive) you may decide that you do not want to do any driving and not exercise that right. If one <u>acts</u> to voluntarily give up the claim, one is said to **waive** the right. For example, you may have the right of way but allow someone else to proceed first. The question of whether one waives a right usually arises when the exercise of that right comes into conflict with something else – in this example, someone else's desire or need to go ahead.

From an ethical point of view, it is crucial for professionals to distinguish between those clients, patients, and students who wish to waive or choose not to exercise some right from those who do not realize that they have the right in question or who do not know how to go about exercising the right. To waive a right, a person must be aware of the right and choose not to exercise it. In some cases others, usually practicing professionals, have an ethical obligation to inform people of their rights. Situations that are unfamiliar to most people or that they do not enter willingly are ones in which people are likely to be ignorant of their rights. Being a patient, being under arrest, or being accused of some wrongdoing are examples of such situations.

Prominent among recently recognized employee rights is the right to know the nature of the occupational hazards to which one is exposed. This right is comparable to a patient's right to be informed of the risks associated with health care and to refuse that care if they wish. Right-to-know legislation requires that workers be informed of the hazards associated with their jobs. The intent of these requirements is to enable workers to make informed choices both about which jobs to accept and how to reduce their risks from hazards they know about.

Recall what makes some claim a <u>moral</u> right: When there is <u>moral</u> justification for some claim, then that person has a <u>moral</u> right. From this definition we see that for a person to have some moral right all that is necessary is that the person's claim be morally justified.

A person's claim (usually) continues to be morally justified even if that person chooses to waive the right in some circumstances. The decision to waive a right in some circumstances does not mean that one waives it in others. For example, in the United States students have a legal right to see records concerning their performance. A student may waive the right to see a particular letter of reference, but the general right remains in force and may be exercised with respect to other material. However, certain rights, such as the legal right to keep others off your land, are forfeited if you do not exercise the right for a given period.

Consider whether a right that is inalienable would always have to be exercised. That a right is inalienable means that the person's claim is always justified, not that the claim must always be pressed. That is, the right does not have to be exercised by the person who has the right, even if it is inalienable.

When, in addition to having a right to do something, a person is morally *required* to do something, the person is said to have an **ethical obligation** or **duty** as well as a right to do the thing in question. An ethical duty or obligation is a moral requirement to follow a certain course of action, that is, to do, or refrain from doing, certain things. It may arise from making an agreement or entering a profession. For example, according to many engineering codes of ethics, engineers have a moral right to raise issues of wrongdoing outside their organizations, but they also have an *obligation* to do so when public health and safety are at stake. [The National Society of Professional

Engineers (NSPE), in their code of ethics, lists thirty-eight professional obligations.]

Returning to the categorization of rights, a term that is often confused with "inalienable" is "absolute." An **absolute right** is a right whose claim can never be outweighed by other moral considerations. The right not to be tortured is widely regarded as an example of an absolute right. This means that no circumstances ethically justify torturing a person.

Some have claimed that rights override, or at least take precedence over, other moral considerations. Philosopher Judith Thomson closely examines this view in her book *The Realm of Rights* and convincingly argues that it is mistaken. One of the counterexamples that she develops in detail is of a person violating another's property right by taking a shortcut and thus trespassing on the other's land because the trespasser is ill and needs medical attention.

The view that rights "trump" other moral considerations is often traced to the contentious and self-assertive character of moral discussion, which is greater in the United States than in other technologically developed democracies. Many thinkers have criticized the tendency to discuss moral life primarily in terms of rights and conflicts among rights and the neglect of responsibilities, virtues, and values realized in community life. Only absolute rights take precedence over other moral considerations.

In contrast to absolute rights, rights whose claims may be outweighed by moral considerations are called **prima facie rights** (from the Latin, "at first face"). Most rights are prima facie rights. For example, the right to travel freely, the right to own a piece of real estate, the right to drive, and the right to be served next (when one has stood in line) are all rights that can be justly overridden under certain circumstances. Of course the circumstances that would qualify can be common for one sort of prima facie right but be rare in others. To say that a right is prima facie rather than absolute is to say that there *might be* other considerations that outweigh the right in a given case.

When the claim of some right is not met, it is common to say that the claim (and the right) is **infringed**. For example, if Alex refuses to give Burt his car keys because Burt is too drunk to drive, Burt's right to drive has been infringed. If a moral wrong is done in infringing a right (i.e., if there are not adequate moral reasons for infringing the right), it is said to be **violated**. If Alex refuses to turn over Burt's car keys simply because Alex is in a bad mood, Alex has violated Burt's right to the use of his property.

An inalienable right need not be an absolute right, because to say that a person has a right that is inalienable only means that there is always moral justification for that person's claim. This does not mean that there couldn't be an even greater moral justification for overriding that claim in some particular situation. Consider the right to travel freely; we regard this as a basic liberty and an inalienable right, but it is only a prima facie right. If people are carrying a dangerous and highly contagious disease, we believe that temporarily overriding their right to travel freely by putting them under quarantine is justified. This example illustrates the point that, in the case of some rights, justice may be best served by overriding (though not disregarding) people's rightful claims.

The same point is illustrated by the fact that we regard it as just for people to be fined or imprisoned in some cases, notwithstanding their inalienable right to liberty and to property. No court could justly take away their right to own property, however, or deny them all liberty by making them slaves, even if they were imprisoned for life. Justice also requires that the amount of the fine and the extent of imprisonment or probation must be in proportion to their offense.

In cases of quarantine or in a case of the police taking over a private automobile in an emergency, the person whose liberty or property is taken has done nothing to deserve forfeiture. For such actions to be just, they must be morally warranted by the importance of the competing considerations – in these cases, protecting the health of the public and coping with the emergency.

Most rights are prima facie rather than absolute. For instance, the right of a (competent) person to refuse medical treatment is another example of a right that is usually regarded as absolute. However, to understand whether a moral wrong has occurred when the claim of some right is not satisfied, it is important to distinguish between different categories of rights.

There is another important distinction between types of rights that cuts across the other distinctions considered so far. On the one hand, some rights require of others only that they not interfere with or restrict the rights-holder. These are called **negative rights** or **liberties**. On the other hand, there are **positive** (or "**affirmative**") **rights**, which are claims to receive something. To respect another's negative right requires only that you not interfere with the person's exercise of the right in question, and not that you provide her with particular opportunities to practice this right. Examples of these negative rights include a person's rights to free speech and to religious expression. In the case of positive rights, it is not enough to leave the rights-holder alone; something must be done for her. Usually some goods or services must be supplied.

Obligations also may be negative or positive (also called "**affirmative**") in the sense just explained for rights. If you pay for the future delivery of an automobile, you have a positive right to the automobile and the seller has a positive obligation to provide it to you. Your right to life, in contrast, is a negative right, that is, everyone else must refrain from killing you – a negative obligation. Your right to life does not impose on others any positive obligation to save your life. Obligations will be discussed at greater length in Parts 3 and 4.

It is commonly held that all people have a right to certain basic necessities, and thus a society that is able to provide them is obliged to do so. Such rights are called **economic rights**, as contrasted with **political rights**. Economic rights are positive human rights. Political rights include physical liberty, or the right to travel freely; freedom of association; and freedom of speech. These examples, which are all freedoms or liberties, suggest that political rights are negative rights, that is, they require only that others not interfere with the rights-holder's activities. However, some political rights, such as the right to vote, require certain services, in this case services that ensure confidentiality and accurate tallying of the vote.

> The relative importance of political as compared with economic rights is widely debated. For example, is a country justified in curtailing political rights to make major improvements in its citizens' economic well-being? Democracies, by their very nature, emphasize political rights. Some democracies, like Sweden, also emphasize economic rights.

Positive rights to health care and to education are often held to be basic human rights. The nature and extent of a right to health care is now widely discussed in the United States. Many people also claim that everyone has the right to a basic education, and indeed public education in the United States is legally mandated for all. Recent legislation affirms the right of people with disabilities to an education in the least restrictive environment possible. This change illustrates how views of the scope of such rights continue to evolve.

> Political scientist Amy Gutmann argues for a basis for legal provision for a basic education, which is different from a human right to education. She argues that the survival of a democracy requires the education of citizens so that they can participate in democratic practices.

The four rights that were taken to be principle human rights at the end of the eighteenth century – the rights of life, liberty, "the pursuit of happiness," and property ownership – were taken to be liberties, not positive rights. Other people generally were regarded as being morally prohibited from interfering with the continuance of another's life, exercise of liberty, pursuit of happiness, or retention of property. They were not morally obligated to save other people's lives, ensure their liberty, promote their happiness, or provide them with property.

In summary, consider what is at issue in the contrast between:

1. alienable and inalienable rights;
2. human rights and special rights;
3. negative rights or liberties and positive rights; and
4. absolute rights and prima facie rights.

The first contrast deals with whether or by what means (e.g., only by forfeiture) the right may be removed from the person; the second with whether the right belongs to all people; the third with whether the claim of the right is to receive something or just to be left alone; and the fourth with whether it can ever be just (morally acceptable) to override the claims of that right.

> The distinction between positive and negative rights helps to clarify some common confusions, for example, one about the right to life and suicide. The view that suicide, or assisting suicide, is wrong requires more than a belief in the right to life. The right to life, even if an absolute right, could nonetheless be waived. Hence, having a right to life does not by itself imply that one also has an obligation to live. Suicide, or requesting or assisting in euthanasia, would be compatible with a right to life, as long as the euthanasia was performed at the uncoerced request of the person whose life was at stake. This point is often obscured because the belief that it is always wrong to take any person's life, including one's own, is often described as a "right to life" position. The view is better described as a belief in the sanctity of life.

MORAL OBLIGATIONS, MORAL RULES, AND MORAL STANDING

Moral obligations and **moral rules** share important characteristics with moral rights. Moral rights, moral obligations, and most moral rules specify what one is morally permitted, forbidden, or required to do without consideration of the consequences of the action – except insofar as these consequences are part of the characterization of the acts themselves;

killing, for example, is an act that results in death. (Obligations and rules, like rights, may be **institutional** or **legal** rather than moral. For example, at many colleges there is an institutional rule obliging all students to see their advisor on or before Registration Day.)

Because rights, obligations, and moral rules concern actions that are permitted, forbidden, or required, they are related notions. Moral constraint on action can be expressed in the language of rights or obligations, as well as that of moral rules. For example, if people have a moral right to refuse medical treatment, then a corresponding moral rule prohibits treating people against their will. Therefore, health care providers all have a professional moral obligation not to perform medical interventions on people without their permission. Moral rights and obligations are subject to further classification, as we saw earlier. For example, rights may be classified as either absolute or prima facie, depending on whether the claims they embody always override other considerations or whether the claims can be overridden by weightier rights and considerations.

Usually statements of obligations specify the acts that are required or forbidden. However, occasionally you will see such statements as this item from an outdated version of the NSPE code of ethics "engineer's have an obligation in their work to ensure the public safety." This provision means that engineers are morally required to ensure the public safety and does not specify what acts they should or should not perform to ensure safety. As we shall see, those obligations are best described as responsibilities. The point of calling them moral obligations is just to say that they are morally required.

Obligations arise from many sources: from one's promises, agreements, and contracts and from one's relationships, debts of gratitude, and roles. Many roles are not chosen, so a person typically has obligations that are not the result of choices, such as the obligation of a son or a daughter. Professional roles are usually at least partially chosen and therefore attendant obligations are in turn chosen.

Often one party's right is matched by an obligation of another party. For example, a client's right to confidentiality is matched by an obligation of the client's consulting engineer to preserve confidentiality.

Rights and obligations have counterpart moral rules. For example, corresponding to the patient's right to refuse treatment and the provider's obligation to give people full information about any risks and obtain their informed consent before subjecting them to hazardous treatment or diagnostic procedures (or making them subjects in any experimental study, hazardous or not) is the rule "Obtain patients' informed consent for hazardous medical interventions or before including them in any experimental study." An engineer's obligation to keep a client's privileged information confidential corresponds to the rule that appears in the codes of ethics of many engineering societies: to keep confidential a client's or employer's business matters.

Recall the earlier definition of negative and positive rights. Would the obligation not to disclose a client's privileged information be a negative or a positive obligation?

As stated, it would count as a negative obligation, because usually it would require only refraining from acts of disclosure. In many circumstances, however, one would actually have to take special precautions to avoid disclosing a client's confidential information – as when one might have to shield a part of a new model from public view. In this case the obligation would require positive action and so would

> Corresponding to people's right to life is a general moral obligation on everyone to refrain from killing other people. The moral requirement can be expressed as a moral rule: Do not kill anyone.
>
> If we say everyone is morally obligated not to kill anyone, we are using "obligation" broadly. Often the term "obligation" is used more narrowly to refer only to moral requirements that arise from the promises, roles, relationships, and memberships of that person – for example, the favors they have received and should repay. Some of the characteristics on which obligations in the narrow sense are based can apply to many people. For example, one might argue that citizens in a democracy have an obligation as well as a right to vote. In the narrow sense of "obligation," only some rights have corresponding obligations. Therefore, although there are "human rights," there are not "human (or universal) obligations" in the narrow sense of obligations.

have the characteristics of a positive obligation. This example illustrates some of the judgments that must be made in applying ethical concepts.

Moral rules or **rules of ethical conduct** specify the acts or course of action required, forbidden, or permitted. In this book I use "rule of practice" or "rule of conduct" to mean specific moral rules that rather precisely delineate the acts or courses of action in question. General rules or admonitions such as "Be honest" or "Treat every person as an end and not as a means" are also moral rules, but they are so general that they tend to be called "basic considerations," "ethical principles," or "fundamental canons." The National Society of Professional Engineers

Some engineering societies, such as the American Society of Mechanical Engineers (ASME) and the American Society of Civil Engineers (ASCE), list both "fundamental principles" and "fundamental canons." (They group specific rules of practice or guidelines under the canons as does the NSPE Code, but they put these rules of practice in a separate document called "Guidelines to Practice.") The principles are general statements, some of which suggest more concern with the well-being of the profession rather than with responsible practice. For example, the first of the ASME fundamental principles is "Engineers uphold and advance the integrity, honor, and dignity of the engineering profession by using their knowledge and skill for the enhancement of human welfare," which thus presents enhancement of human welfare as a means to the end of enhancing the profession.

(NSPE) in their Code of Ethics classify as "fundamental canons" such general imperatives as, "that engineers shall perform services only in areas of their competence or that they shall issue public statements only in an objective and truthful manner." In contrast, the NSPE calls specific moral rules (grouped under each of the fundamental canons), "rules of practice." For example, under the canon enjoining truthfulness and objectivity are two rules: "Engineers may express publicly technical opinions that are founded upon knowledge of the facts and competence in the subject matter" – a rule that supports an engineer speaking out – and "Engineers shall issue no statements, criticisms or arguments on technical matters which are inspired or paid for by interested parties, unless they have prefaced their comments by explicitly identifying the interested parties on whose behalf they are speaking, and by revealing the existence of any interest the engineers may have in the matters," a rule requiring engineers to disclose any conflict of interest.

Other codes use slightly different language, which is partly a function of the way in which the code is organized. The Code of Ethics of The Institution of Engineers, Australia (IEA) states three general cardinal principles, such as "to respect the inherent dignity of the individual" (which they say should apply in personal as well as professional life), and states nine "tenets," which are rather general rules for the ethical conduct of professional practice. For example, "members shall apply their skill and knowledge in the interest of their employer or client for whom they shall act as faithful agents or advisers, without compromising the welfare, health, and safety of the community." The IEA code also offers clarification of some conceptual points. For example,

> Members should understand the distinction between working in an area of competence and working competently. Working in an area of competence requires members to operate within their qualifications and experience. Working competently requires sound judgment.
>
> – Code of Ethics of The Institution of Engineers, Australia (IEA)

Some writers state principles in the form of obligations or general rules of action, but you will also see references to, for example, the "principle of honesty" or "the principle of respect for persons." It is easy enough to put such principles in the form of rules, however, assuming one understands what the principle says. The principle of honesty clearly translates as the basic rule: Be honest. "The principle of respect for persons" is a somewhat jargoned way of saying "Respect other people's right of self-determination, that is, their right to manage their own lives." The cultural, religious, and intellectual heritage of the person or group formulating or naming the principle influences the name or formulation. (For that reason you are encouraged to phrase principles in terms that connect the present subject matter to the moral categories you already recognize. The disciplinary jargon that some scholars use to formulate ethical principles often reflect their positions in scholarly disputes, and those disputes may not be relevant to deepening your own understanding of ethical problems in engineering practice or research.)

Specific rules of practice may evolve into ethical principles in much the way that specific empirical scientific laws (laws based on observed regularities, rather than derivation from other laws) evolve into principles. In both cases we give the name "principle" to those relationships that we regard as fundamental and the truth of which we assume in making other observations and inferences. The French mathematician and philosopher, Henri Poincaré,

> As philosopher Alasdair MacIntyre argues, moral rules and principles are learned and formulated in the context of situations to which they apply, rather than being known abstractly, like principles of logic. He says a "moral principle or rule is one which remains rationally undefeated through time, surviving a wide range of challenges and objections, perhaps undergoing limited reformations or changes in how it is understood, but retaining its basic identity through the history of its applications. In so surviving and enduring it meets the highest rational standard."[7]

described this process for many physical laws and principles. For example, Newton's second law, $f = ma$, was an empirical law when Newton first proposed it, but it quickly evolved into a principle and came to function so that now it functions virtually as a *definition* of force. On the ethical side, people often speak of "the principle of informed consent" because the rule of informed consent has become a basic element in ethical reasoning about many ethical matters other than health care and human experimentation. (Despite the similarity of the relation of physical principles to empirical laws and ethical principles to rules of ethical practices, rules of ethical practice are different from scientific laws in other respects.)

Together with the formulation of related principles, the character of the situations for which the rule is offered influence the formulation of that rule. The influence of problem situations on rules of practice will be amply illustrated in Chapter 2 when we further examine some rules in some professional societies' codes of ethics.

To summarize: A moral rule is an ethical standard in the form of a rule of behavior. An ethical principle or fundamental canon is a general moral consideration that provides the framework for more specific rules of practice. Therefore, the term "moral rule" may apply both to ethical principles and rules of ethical practice.

There is a legal counterpart to the notion of a moral obligation. A legal obligation is a legal required behavior. Negative legal rights, which are the counterpart of moral liberties, specifically forbid acts of interference or harm. Positive legal rights are legally warranted claims to *receive* something – for example, a due process proceeding or compensation for past injury. Although many legal rights are also moral rights, the two notions are not coextensive. For example, people have a moral obligation to keep their promises, although not all promises are covered by legal statute. Furthermore, as the laws upholding slavery illustrate, legal and moral obligations may conflict.

Although any obligation has a corresponding moral rule, not all obligations or moral rules have corresponding rights. There are moral rules that apply to the behavior of moral agents toward beings who, although their welfare must be considered, are not the sort of beings that have rights. It is commonly held that moral obligations and moral rules apply to the treatment of human corpses and to nonhuman animals. These are examples of moral obligations and moral rules in the absence of corresponding rights. Consider the treatment of human corpses. Some religions hold that the treatment of corpses affects the person whose body it was, but most people recognize the moral rule that they ought to treat human corpses with respect even if they do subscribe to such a belief. A variety of nonreligious reasons are given for believing that people should treat corpses with respect, including that disrespecting corpses is likely to increase callousness toward the living.

The question of moral constraint on the treatment of human corpses was discussed with practical application to product development a few years ago when it was decided to resume using human cadavers in auto safety test crashes. Treatment of corpses is also of practical importance in setting practices of teaching hospitals, which sometimes allow student physicians to practice medical procedures on corpses before rigor mortis sets in. This practice affords prospective doctors the opportunity to increase their proficiency before they apply medical procedures to living patients. Laws requiring the consent of the family for any procedures done to the corpse are common and reflect the repugnance with which most people view the instrumental use of corpses. However, this legal restraint is commonly circumvented by the ploy of delay in pronouncing the patient dead.

The broad ascription of rights to beings who do not make reflective choices has become widespread in the United States in the last few decades along with heightened concern about the welfare of nonhuman animals. However, whether someone ascribes rights to nonhuman animals does not fully determine the person's view about how such animals ought, ethically speaking, to be treated. In practice, there is only a very general tendency for those who hold that animals have rights to think that animals should be treated much as we treat persons. Many who are reluctant to ascribe rights to nonhuman animals do recognize obligations of people toward them. As already mentioned, a moral prohibition on cruelty to animals is widely recognized and is backed by some laws.

> It is not entirely clear what it means to treat a nonhuman the way one would treat a person. For example, if you come across an injured wild rabbit and you are a conscientious animal rightist, ought you leave the animal to its natural devices or get veterinary help for it? It turns out that there are some who call themselves animal rightists who think people should not interfere with animals even to the extent of having pets. Other animal rightists believe you should show the same concern for relieving animals' pain and suffering as you would for a human.

Since the strictness with which one uses the term "rights" does not settle questions regarding the obligations of moral agents toward beings who are not moral agents, we must look further.

The question of the moral limits on experimentation with animals is of particular importance for science. Some scientists burn and maim animals to devise treatments for burned and maimed people. Furthermore, because anesthesia and analgesics would interfere with some of these experiments, the animals are not given anything for their pain. Just because these acts are experiments does not decide the question of whether they are also acts of cruelty. Therefore we must ask whether we should view these acts as cruel; whether such cruelty constitutes a violation of moral rules or obligations toward animals; and whether this violation can be justified.

What considerations are relevant to determining whether it is morally justifiable to do experimentation with animals? The first consideration is what happens to the animal, whether it is disabled, killed, or caused pain. Beyond that, it is morally relevant whether the obligation not to cause animals severe pain when their own welfare is not promoted is an absolute or only a prima facie obligation. If it is prima facie, justification would depend on the relative strength of the countervailing considerations, such as nature of any benefit to people achieved through the experiment.

Many who wish to ascribe rights to nonhuman animals contend that their well-being is important in itself, regardless whether their well-being contributes to the well-being of humans. When the welfare of some creature must be considered for its own sake, it is said to have **moral standing** or intrinsic moral worth. To say that some group of beings have moral standing does not decide the question of whether they have *the same* moral standing as people and thus have "human" rights, but only that the welfare of such beings must be considered for its own sake. Consideration of moral standing does function like consideration of rights in that consideration of obligations toward a being with moral standing also constrains how far a person may go in seeking to improve an outcome. If, for example, the moral standing of some beings, let us say beings without a right to life to differentiate them from people, makes it wrong to subject them to great pain, then it is prima facie wrong to experiment on them without anesthesia, and anyone who claimed such experiments were justified would have the burden of showing that it was.

The history of ethical thought shows that those in power have often recognized the moral claims of others who were dissimilar to themselves. The human rights of many people have been ignored because of their race, class, or gender. That behavior is now described as racism, classism, or sexism and is understood as unwarranted preferential treatment of the race, class, or gender in power. If one claims that humans are the only group with moral standing, this looks suspiciously like a new un-warranted bias, a bias in favor of the human species – what Peter Singer has called "speciesism." To show that the claim of moral standing (or absence of moral standing) of members of other species is more than an arbitrary exercise of prejudice on the part of humankind, one must show that distinctions in moral standing are based on morally relevant features of the beings in question. Another way of expressing the view that a being has moral standing is to say that its well-being (or at least some aspects of it) is of value in itself and is not merely a means to other desirable ends.

Those who claim that nonhumans have moral standing and those who say that animals have rights often agree on what they believe is morally required in the treatment of animals. Both groups tend to disagree with people who are concerned about the treatment of nonhuman animals only if (and to the extent that) humans are affected by that treatment. However, even those who are concerned about the treatment of non-human animals only to the extent that such treatment has an effect on human well-being may object to cruelty toward nonhuman animals on the grounds that cruelty is a moral vice and thus cruelty to animals should be avoided as morally corrupting. Therefore, denying moral standing to some beings does not thereby commit one to the view that "anything goes" with regard to their treatment.

As Robert Proctor points out, the Nazis were staunch defenders of the view that it was wrong to victimize healthy specimens of other species by using them for scientific experiments and that it was morally preferable to use as subjects "defective" humans.[8] This example illustrates the point that cruelty to one group can easily coexist and even seek justification in compassion toward another.

PART 3. MORAL CHARACTER AND RESPONSIBILITY

VIRTUES AND VICES

In contrast to moral rights, moral rules, and moral obligations, traits of **moral character**, or **virtues** and **vices**, such as honesty, kindness, cow-ardice, and responsibility, are characteristics of people, rather than their acts. As Alasdair MacIntyre has argued, traits of virtues are essential in the development of complex cooperative activities. Social practices achieve ends and produce results for which their practitioners may receive what MacIntyre calls "external" rewards: pay, fame, or career advancement. For example, epidemiological research produces new knowledge about diseases in populations, for which researchers may receive external rewards in addition to knowing that they have succeeded in advancing knowledge. In addition to developing certain skills, engaging in a practice also develops certain virtues in its practitioners. Practicing epidemiological research, for example, develops not only research skills, but also virtues such as patience, thoroughness, and diligence. MacIntyre calls these virtues "internal" goods, or rewards of the practice, because they are achieved quite apart from whether the research yields any particularly notable results or advances researchers' careers. The character traits considered desirable and that are, therefore, called virtues vary somewhat with sphere of activity and the relative importance that specific activities are accorded in particular cultures. For example, the intellectual self-discipline required to rigorously test hypotheses in engineering and other scientific fields may not be an important character trait in parenting a young child. Nonetheless, many scholars agree that some virtues, such as honesty and courage, are necessary to the successful conduct of all or most social practices.

To understand a person's character, one must understand the whole configuration of ethically relevant considerations that influenced his actions. Knowing that a person often broke the law might lead one to conclude the person was dishonest. However, if the individual habitually hid people from unjust persecution by a tyrannical government, then the person could well have been an honest person in circumstances that justified lying to law enforcement officials. If a person's apparent bravery and willingness to risk his life in battle derived mainly from an obsession with killing and maiming people, then the quality would not be the virtue of bravery, but merely the expression of a character defect in a socially acceptable way.

Although the concept of **moral integrity** is central to the assessments of character, it does not comprise an additional character trait. Roughly, "moral integrity" is the ethical coherence of a person's life and actions. Honesty and consistency characterized by the absence of hypocrisy or betrayal is part of the notion. People's values may be expected to develop over the course of their lives. Hence moral integrity is not simple persistence in maintaining value commitments. The coherence of a person's life is a narrative coherence, that is, to understand a person's character and moral integrity, one needs to understand the place of values or ideals as they develop throughout a person's life.

A loss of integrity can be forced upon a person. One example is Sophie, in the book and film *Sophie's Choice*. Nazi guards force a true dilemma on Sophie: She is required to choose, in the presence of her two children, which of the two is to be killed on pain of having them both killed. Being forced to send one child to her death is fatal to Sophie's moral integrity and sense of self. This is an extreme case, but it illustrates that circumstances as well as personal resolve are factors in maintaining moral integrity.

A more common situation is one in which all of the obvious responses a person can make threaten to betray some relationship or trust. This can happen when someone is called upon to make a grave healthcare decision on behalf of a family member, one in which the decision maker is unprepared to make in a way that he feels the ill person would have wanted.

In developing or reviewing policies and practices in a work situation, it is important to be alert to mismatches – or even conflicts – between the skills and virtues of key actors and the skills and virtues that others need in those key actors. Even well-meaning people can respond badly when they have not thought through how they will fulfill potentially conflicting responsibilities simultaneously. For example, a devotion to the progress of scientific research might interfere with a health care provider's responsibility to secure the best health outcome for his patients or with an engineering faculty member's oversight of her graduate student's education.

ETHICAL RESPONSIBILITY AND OFFICIAL RESPONSIBILITY

For someone to have a **moral responsibility** for some matter means that the person must exercise judgment and care to achieve or maintain a desirable state of affairs. Notice that we speak of people reaching "an age of responsibility" or "age of discretion," indicating that although children may follow moral rules, something more is required in terms of cognitive ability or matured judgment to exercise responsibility appropriately.

The moral sense of responsibility, in which one undertakes to achieve some future state of affairs or maintain some present one, should not be confused with the *causal sense* of responsibility for some existing or past state of affairs. (Recall the example of the storm that was said to be "responsible for" deaths and property damage. This was causal responsibility not moral responsibility. Attribution of responsibility to the storm means only that the storm caused particular outcomes. As we saw, storms do not have moral responsibilities and are neither responsible nor irresponsible in the moral sense. They are causal but not moral agents, so their actions are not subject to moral evaluation.) Moral responsibilities of a moral agent may derive from their causal responsibilities, however. If a person has caused a difficulty, there is reason to think that the person has some moral responsibility for remedying the resulting situation. If you break something, you have some responsibility for fixing it or for cleaning up the mess and replacing it. However, people often find themselves faced with a responsibility not of their own making. If an infant or young child breaks something, someone else must clean it up. There is much discussion of the fact that if pollution of the environment is not adequately addressed in one generation, subsequent generations find themselves responsible for cleaning up the contaminants that another has left.

Characteristically, the achievement of the desired outcome involves some exercise of discretion or judgment. This is what distinguishes a responsibility from other moral requirements. An obligation or duty specifies what acts a person is required to perform or refrain from performing. Notice that this difference is reflected in the difference between the expression "responsible for (some end)" – such as responsible for the safety of some device or responsible for the welfare of some person – as contrasted with "obligated to *do* or refrain from doing certain things." Often what the obligation states rather specifically are the acts one is expected to perform or refrain from performing. Contrast a professional's responsibility for the well-being of her clients with a professional's duty or obligation to be truthful about her qualifications or anyone's obligation to refrain from assaulting people.

The relation between an obligation and a responsibility is actually somewhat more complex, and the two overlap. To see how, notice that the Code of Ethics of the National Society of Professional Engineers after saying that engineers shall "Hold paramount [that is, take as their primary responsibility] the safety, health and welfare of the public in the performance of their professional duties" goes on to say "Engineers shall at all times recognize that their primary obligation is to protect the safety, health, property and welfare of the public." Consider the obligation stated in this last passage. It is stated in the form of an obligation to *do* something (in this case protect the safety). The obligation is stated, not in terms of the precise acts one is to perform or refrain from performing but in terms of what one is to achieve, namely preservation of the safety, etc. of the public. This illustrates that an obligation may be specified in terms of what one is to *achieve* rather than what acts one is

INTRODUCTION TO ETHICAL CONCEPTS

expected to perform, in which case the obligation is interdefinable with a responsibility. The obligation to refrain from taking bribes specifies what *acts* are forbidden, namely the offering of payments or inducements to someone in a position of trust to get them to do something for the bribe payer to which the bribe payer is not entitled. In contrast the obligation to protect the public safety specifies what you are to *achieve* rather than what acts you are to perform or refrain from performing. Therefore, the obligation to protect safety may be expressed as a responsibility *for* public safety. The terms "obligation" and "duty" can be used to describe some matters of responsibility.

Further confusion may be caused by the fact that the term responsibility is sometimes used as a synonym for obligation, so that one may say, for example, "It is your responsibility to back up the computer files before you leave." To avoid confusion in this book the term "responsibility" will never be used as a synonym for obligation, that is, it will never be used in the form "responsibility to perform some act." It will be used exclusively in the form "responsibility for some outcome or state of affairs to be achieved."

Moral responsibilities derive either from one's interpersonal relationship to a person whose welfare is in question or from the special knowledge one possesses, such as professional knowledge that is crucial to an aspect of another's well-being. Examples of the first sort include the responsibility of one friend for another and of a parent for a child. Notice that a person can have this first kind of responsibility without having any particular knowledge that helps him fulfill the responsibility. Examples of the second sort are the responsibility of a health practitioner to stop and give aid to an injured person who may be a stranger and the responsibility of an engineer to ensure public safety and thus safeguard many individuals whom the engineer will never meet. One person's responsibility for another's welfare may combine both elements. For example, a healthcare practitioner may have a significant personal relationship with a patient who also is dependent on the practitioner's knowledge for adequate care. Since few relationships and knowledge are shared by everyone, most moral responsibilities are special moral responsibilities, that is, they belong to some people and not others. There is no generally accepted category of "human responsibilities" as there is human rights or (by derivation from human rights) human obligations.

Professional responsibility is the most common type of moral responsibility that arises from the special knowledge a person possesses. Mastery of a special body of advanced knowledge, particularly knowledge that bears directly on the well-being of others, distinguishes professions from other occupations. In modern times it is simply not possible for a person to master all the knowledge relevant even to one's own well-being. Because society looks to members of a given profession to master and develop knowledge in a particular area, the members of a profession bear special moral responsibilities in the use of the special knowledge vested in them. A state environmental protection division employs an environmental engineer to decide whether plans for construction of a power plant meet the regulation requirements of the Clean Air Act, that is, whether the plans provide sufficiently for reduction of such pollutants as sulfur dioxide and the nitrous oxides and thus whether a building permit should be issued. Engineering knowledge is required to be able to make this assessment.

Although some moral demands on professionals are adequately expressible in rules of conduct that specify what acts are permissible, obligatory, or prohibited, there is more to acting responsibly than following rules. A good consulting engineer not only shuns bribery, checks plans before signing off on them, and the like, but also must exercise judgment and discretion to provide a design or product that is safe and of high quality. Moral agents in general, and professionals in particular, must decide what to do to best achieve good outcomes in matters entrusted to their care.

Not only does responsible behavior require more than the performing specified acts, but the person with the responsibility need not be the one to perform the acts that are necessary; this person need only see that someone else does. Thus the question "Who will be responsible for the lead screening program?" does not ask who will do the screening tests, but rather who will see that the program is carried out.

Now consider the differences between a moral responsibility and an **official responsibility** – that is, a responsibility that someone is charged to carry out as part of one's assigned duties. The description of a job or office specifies some of one's official responsibilities. One could argue that there is a prima facie moral obligation to keep one's promises, and when one takes a job, one implicitly promises to perform the obligations or "duties" that go with that job. One is, therefore, morally obliged to fulfill those responsibilities because one has promised to do so. In this way, official obligations and responsibilities, then, can become moral responsibilities and moral obligations to the extent that one freely takes on a job or office. Moral responsibility, however, does not reduce to official responsibility. Indeed some official responsibility or obligation may be immoral. "I was just doing my job," or "I was just doing what I was told" is not a generally valid excuse for unethical behavior on the part of an adult.

Corresponding to the notion of moral responsibility is the notion of legal responsibility. A legal responsibility may arise in either of two ways: as a moral responsibility that is legally recognized and enforced or as a legally mandated official responsibility. An example of the latter is the legal responsibility for deciding whether to move some community members from their homes to prevent their further exposure to toxic contamination, a legal responsibility that may be part of the job of a public official.

The notion of official responsibility is central to the attribution of decisions to organizations rather than to the people in them. For example, people may say that the Ford Motor Company made the decision to rush the Pinto into production, rather than that particular people, such as Lee Iacoca, then president of Ford, made the decision. This way of thinking about decisions turns on the idea that an organization is a "decision-making structure" and that the actual person or people who make a decision carry out their official responsibilities and obligations according to the values and criteria handed down by the company. Organizational values specify all of the goals to be achieved. On this view the technical skills and scope of authority specify the scope of actions that the agent is to take in achieving those organizational goals. The agent's own values or the values of the agent's profession, religion, or culture are all assumed to be irrelevant to what the agent will do in "doing one's job." Therefore, in this model doing one's job is

According to Herbert A. Simon's model of organizational behavior, people in formal organizations ideally make the decisions delegated to them on the basis of the organization's interests and values, rather than on the basis of the values they themselves hold. Simon presents this model, not as a description of how administrators make decisions, but of how they ought to. That makes Simon's model what we call a "normative" theory, rather than a descriptive theory, and others have challenged it. This model of organizational behavior treats one competent person as completely substitutable by any other who comes to occupy the same position in the organization; that is, any agent in a given position would have the same official responsibilities, and any competent person in that position would make essentially the same decision.

unaffected by the character and values of the person doing the job. Any decisions that a person makes in his or her official capacity are attributable to the organization rather than the individual.

As John Ladd has argued, *official* responsibilities differ significantly from moral responsibilities in that they attach to job categories and impersonal roles rather than to particular people in particular circumstances with histories and human relationships that are unique to them. [9]

The scope of one's official responsibilities are specified by one's position, and one's job description, apart from one's own insights into the situation. One person's official responsibilities exclude another's. This exclusionary feature makes official responsibility quite unlike moral responsibility. Two friends of the same person may both have a moral responsibility to see that the person does not drive while intoxicated, for example.

If a supervisor were to say to an engineer, "It is not your job to think about safety questions," this might be true as a statement about official responsibilities but would not mean that the engineer lacked any moral responsibility for raising safety concerns. Although a person's job description may not include some matter, he or she may have a moral responsibility in that matter, especially if it is a responsibility of their profession.

Moral responsibility, unlike official responsibility, cannot be simply transferred to someone else. This feature of moral responsibility is expressed by saying that it is not alienable. If an engineer in charge of a project assigns to another member of the team the responsibility to make certain safety checks and the subordinate fails to do so, the engineer in charge will bear some responsibility for the failure, especially if the engineer in charge had reason to know that the subordinate was not reliable or did not have the relevant competence.

Consider the following case based on real-life events and reviewed by the Board of Ethical Review (BER) of the National Society of Professional Engineers (NSPE):

THE RESPONSIBILITY FOR SAFETY AND THE OBLIGATION TO PRESERVE CLIENT CONFIDENTIALITY[www]

The owners of an apartment building are sued by their tenants to force them to repair defects that result in many annoyances for the tenants. The owner's attorney hires Lyle, a structural engineer, to inspect the building and testify for the owner. Lyle discovers serious structural problems in the building that are an immediate threat to the tenants' safety. These problems were not mentioned in the tenants' suit. Lyle reports this information to the attorney who tells Lyle to keep this information confidential because it could affect the lawsuit. Lyle complies with the attorney's decision.

—adapted from NSPE Board of Ethical Review Case 90-5.[www]

What, if anything, might Lyle have done other than keep this information confidential? Which, if any, of those actions would have better fulfilled Lyle's responsibilities as an engineer?

What other information may be needed to make this decision?

The question that the BER explicitly addresses in its discussion is whether certain actions of engineers described in the case are "ethical or unethical," and their decisions are based solely on applicable provisions in the NSPE code of ethics. The reasoning behind these simple binary judgments is what makes them interesting. The discussion of these cases reflects the norms of ethical practice put forward by this professional society. The cases that come to the attention of the NSPE's Board of Ethical Review are predominantly ones in which one licensed engineer has a complaint about another.

There is some danger that in emphasizing the professional responsibility to work for the well-being of a client – rather than just emphasizing the rights of the client – we encourage **paternalism** on the part of the professional. Paternalism derives from the Latin word for father (*pater*). Acting like a parent toward those who are not your children may or may not be justified in particular circumstances. An act of paternalism may be roughly defined (following Gert and Culver) as infringing a moral rule of conduct toward someone or infringing that person's rights (such as the right of self-determination) for what the agent believes is that person's own benefit.

The question of paternalism often arises in medicine and health care with respect to the treatment of patients. Because many engineers in industry must protect the safety and health of anonymous members of the public rather than identified clients, and because they usually do not occupy positions of greater power than do the clients they have, paternalism is not a frequently discussed topic for engineers in industry. However, even for such engineers in industry, the issue of paternalism can arise in connection with "idiot-proofing," as we shall see in

Chapter 3. Issues of paternalism often do arise for engineers and scientists in connection relationships among coworkers and students.

THE CASE OF MEAGER FIRST-AID SUPPLIES

The first-aid kits in some of the university teaching laboratories contain only bandaids. When some members of the engineering faculty tried to have more adequate supplies put into the kits, they were told that if the kits contained more supplies, those supplies might be misused in a way that would cause injury. Anyone who needs more than bandaids, they were told, should go to the health center for treatment.

This example illustrates that if one is determined not to put anything in peoples' hands with which they might harm themselves, they will not be able to do themselves much good either.

To say that some act counts as paternalism does not yet tell us whether it is justified or unjustified paternalism. However, acts of paternalism do need justification, because they involve infringement of moral rules; the burden of proof is on the side of those who claim that a given act of paternalism is morally acceptable. Furthermore, the responsibilities of a professional to look out for a client's welfare in the area of the professional's expertise need not conflict with any of the client's rights, especially if the professional explains the pros and cons of the situation to the client rather than simply making a judgment that is left unexplained.

Because paternalism involves the infringement of moral rules, it needs justification. If the rule infringed in a given case was not an absolute moral rule, then other moral considerations may show that the act of paternalism was, on balance, *justified*; that is, it was right to do it in those circumstances.

TRUST AND RESPONSIBILITY

Trust of many sorts is necessary for ordinary life: trust of technology, trust of institutions, trust of other individuals. Without trust there can be no cooperative activities and thus no life in a community or society. (Cooperative activities include many that are also competitive. Competitive sports is a handy example of a competitive activity in which there are standards of "fair competition" to be mutually upheld.) **Trust** is confident reliance; confidence and reliance do not always go together. We may rely on someone or something, trusting that the thing or person in question will perform as needed and expected. However, we may also rely on people or things even where we have good reason not to trust them. If I am told that my well may have been contaminated with toxic substances, then I will stop using water from the well only if or to the extent that I have another source of water available. On the other hand we may have great confidence in something – say that the automobile of the president of General Motors is in good repair – but unless we can in some way rely on this fact, we do not trust in it.

As Annette Baier has argued, trust does not always have an ethically sound basis. Someone may trust another whom she has successfully threatened or otherwise coerced into doing her bidding. Baier's general account of the morality of trust illuminates the strong relation between the trust *worthy* and the true. A trust relationship according to Baier is decent if, or to the extent that, it stands the test of disclosure of the basis for each party's trust. For example, suppose one party trusts the other to perform as needed only because the truster believes the trusted to be too timid or unimaginative to do otherwise. Or suppose the trusted fulfills the truster's expectations only because he fears detection and punishment. Disclosure of these premises will undermine the trust relationship. Knowing the truth will give the trusted person an incentive to prove the truster wrong, or give the truster the knowledge that if undetected defection or betrayal becomes feasible, the trusted will likely defect or betray. Telling the truth about the basis for trust is an operational test of whether the trust is rooted in trustworthiness and a confidence in the other's trustworthiness. If the trust relationship cannot withstand having the truth told about it, it is corrupt.

Although explicit discussion of moral trustworthiness is relatively recent, both students of professional ethics and the philosophy of technology have given considerable attention to the concept of responsibility. Being trust *worthy* is key to acting responsibly in a professional capacity or being a responsible person in the virtue sense of "responsible." Therefore, the literature on responsibility, which has been extensive in recent discussions of professional ethics, provides at least an implicit discussion of many aspects of the morality of trust in this professional practice.

Since being a responsible person means being able to take responsibility for one's own actions, it is closely connected to rights of self-determination. Foremost among these are rights to one's person – body and mind – although property rights have often been argued (e.g., by

John Locke in the seventeenth century) to derive from one's rights to one's body and the fruits of one's labor. If a person's rights to his body and mind are not respected, the person's actions are not his own in an important sense. If, for example, people were to be drugged, it would effectively undercut the fulfillment of their moral responsibilities, personal and professional, and so undercut the rest of moral life.

PART 4. PRIVACY, CONFIDENTIALITY, INTELLECTUAL PROPERTY, AND THE LAW

PRIVACY AND CONFIDENTIALITY

As we saw earlier, human rights reflect moral claims without which it is difficult for people to act as moral agents. Often included among these basic rights are rights of **privacy**. However, privacy is a notion that receives more attention in individualistic cultures than in others. In fact, as was mentioned earlier, some cultures do not have a word for this notion.

The claim to privacy finds moral justification in the recognition that in order to function as moral agents, people need to have control over matters that intimately relate to them. Therefore, privacy is closely connected to the right of self-determination. What one person is expected to do to respect another's privacy varies from culture to culture. In some contemporary cultures, parents oversee their children's affairs much more closely than in others. In a traditional Chinese culture, it is expected that parents will do such things as read the mail addressed to their adolescent children as part of their responsible oversight of the children. In Anglo-American culture, such acts would be viewed as intrusions of the adolescent's privacy.

Many questions of employee privacy arise in the workplace, including privacy of e-mail and telephone and fax communications and employee drug testing. Scientific discoveries and technological innovations have made possible the acquisition of new types of sensitive information, such as genetic information. Computers and other information and communications technology make it possible to collect, correlate, and transmit quantities of personal information in ways that previously were impossible.

Three types of privacy are commonly distinguished: physical, informational, and decisional (Allen, 1995, p. 2065f). In addition, philosopher and legal theorist Anita Allen distinguishes dispositional privacy (Allen,

1987, pp. 15–17). Physical privacy is a restriction on the ability of others to experience a person through one or more of the five senses; informational privacy is a restriction on facts about the person that are unknown or unknowable; and decisional privacy is the exclusion of others from decisions, such as health care decisions or marital decisions, made by the person and his group of intimates. Finally, dispositional privacy restricts what others may legitimately know of another person's states of mind.

As an example of how dispositional privacy differs from informational privacy, consider the Lotus Corporation's proposal to sell their Market-Place Data Base, a database that contained extensive information about the consumption patterns of large numbers of people. This was widely criticized as an intrusion of the privacy of those profiled, even though the items of information aggregated were not the sort of information normally considered private.[10]

The aggregated data offered by Lotus gave, and was intended to give, dispositional access to people's states of mind, specifically their consumer preferences, whereas the component items of information did not. Madison Powers[11] rejects Allen's distinction and argues for reducing all three types of inaccessibility to informational inaccessibility. However, the Lotus example speaks for the greater adequacy of Allen's scheme: If knowing a person's dispositions is a species of privacy invasion, then the aggregation of nonprivate facts can result in a privacy invasion, if taken together the facts do give evidence of dispositions.

Once the questions of appropriate levels of privacy protection have been determined, the question of how to practically ensure that level of privacy becomes a matter of **security**. The security of a system is the extent of protection afforded against some unwanted occurrence, such as the invasion of privacy, theft, the corruption of information, and physical damage.

One strategy to preserve privacy involves secrecy. Another is to keep sensitive information confidential. Confidential information is information that may be shared, but only within a restricted group, usually those involved in some joint task who have a need to know the information. For example, information in a medical record is confidential information that is used by health workers who are bound not to disclose the information to outsiders except for legitimate purposes such as insurance reimbursement. Should someone, say a biomedical engineer, wish to report on a clinical case at a conference, that person must first remove identifying information about the patient unless the patient explicitly agrees to be identified.

INTELLECTUAL PROPERTY

In engineering practice a common reason for holding information confidential is to preserve rights to intellectual property. For example, if information involving a trade secret or a company's business plan is disclosed, then this disclosure could give competitors an advantage. A **trade secret** is a device, method, or formula used in one's trade or business that gives one an advantage over the competition and which must be kept secret to preserve that advantage. The formula for Coca-Cola is a trade secret. If the holder of a trade secret takes sufficient precautions to keep the secret secure, then the courts of the various states will protect it as a form of intellectual property. Others will be legally restricted from wrongfully taking, using, or disclosing the trade secret. Disclosure of confidential information is an example of such wrongful action. Acceptable means of learning a trade secret include reinventing it or learning it through reverse engineering of a purchased product.

Secrecy is not a very good way of restricting use of one's intellectual creations if the secret can be readily learned by acceptable means. Copyrights, patents, and trademarks give legal protections to fully disclosed creations. A clause of the U.S. Constitution provides for encouraging the development of science and the useful arts by granting to authors and inventors a time-limited exclusive right to their writings, discoveries, and inventions, and federal statutes specify the rights.

A **patent** is a special, alienable legal right granted by the government to make, use, and sell, or at least (in case there are other patents that the new patent holder's use of her patent would infringe) to bar others from making, using, or selling a device, design, or type of plant that one has created. In the United States the time period is twenty years for useful devices and fourteen years for designs. In taking out a patent one makes the nature and details of one's invention plain or "patent." In the United States, to be eligible for a patent an application must be initiated within one calendar year of "public disclosure" of the idea. The European and Japanese patent laws require application prior to public disclosure. However, most countries honor the patents taken out in other countries. Although the inventor(s) name(s) always appear on patents, the property rights may be assigned to others. This illustrates the point that the rights we are discussing are a kind of property; they are special alienable rights. These rights of ownership are distinct from the credit that goes to an author, composer, or inventor. Some codes of ethics of professional societies require both general crediting of intellectual work and honoring property rights embodied in patents, copyrights, and trademarks. For example, the NSPE Code of Ethics specifies the following among an engineer's professional obligations:

> Engineers shall give credit for engineering work to those to whom credit is due, and will recognize the proprietary interests of others.

Credit is not the same as a property right, however.

To patent a device one must prove that it is useful, novel, and not obvious. Therefore not all inventions are patentable. This is one reason for holding some inventions secret rather than patenting them. Another is that trade secrets have no expiration date as does patent protection.

To establish that a patent is valid, it must often survive a court challenge. Obtaining a patent constitutes a significant expense, and fighting or defending one in court can be extremely costly, often costing a million dollars or more in legal fees. This is a third reason that many inventors either try to protect their intellectual property in other ways or hold the patent jointly with some organization that has a legal staff. A **trademark** is an officially registered name, symbol, or representation, the use of which in commerce is legally restricted to its owner.

A **copyright** is a legal right to exclusive publication, production, sale, or distribution of some work. The proprietary interest that is protected by the copyright is the "expression," not the idea. Ideas cannot be copyrighted. (Taking credit for another's idea is plagiarism, so copyright protection is not the legal equivalent of a prohibition of plagiarism.) Copyright protection of a work under the U.S. 1976 Copyright Act begins as soon as a work exists in a concrete form and remains in effect until fifty years after the death of the author. A copyright is most commonly held by the author, composer, or publisher of a work, but it may be assigned to others or inherited.

Many legal notions were discussed in this section, but it is important to distinguish legal from ethical considerations. The next and final section of this chapter will consider the ethical standing of the law and various forms of conscientious refusal where no legal questions are at stake.

ETHICS, CONSCIENCE, AND THE LAW

In the earlier discussion of moral rights I noted that moral rights and legal rights must be distinguished, that a person may have a legal right to do something, but not a moral right to do it, and some laws are unjust. What does this mean about the moral force of law?

"Those who enjoy sausage and respect the law should never see either being made." The implication of this saying, attributed to Bismarck, is that seeing the influence of particular interests on the formulation of legislation can lead one to doubt the moral authority of the law. Most societies attempt to have laws that are just and morally sound, notwith-standing the particular influences on a given piece of legislation. Many individual laws are neither morally sound nor unsound, just nor unjust considered by themselves. For example, requiring everyone to drive on the right side of the road is not morally superior to requiring everyone to drive on the left side of the road, although it certainly is prudent to require everyone to drive on the *same* side of the road. The entire system of making and applying laws in legal judgments is supposed to be just, though not infallible. At least in most functioning democracies, law has moral authority even though individual laws may have no moral significance, may be poorly written (so they do not accomplish what they were intended to), or even without justification may favor some person or group. The burden of proof is on any person who claims that some law is unjust or that one ought not obey it. Ethically speaking, people are expected to be law-abiding except where they can show good reasons for thinking some law unjust or immoral.

Openly breaking a law to test its constitutionality is not disrespecting the law, but using part of the legal system, the court system, as quality control on the legislation, so to speak. Breaking a law so as to test it not by a constitutional standard but by some other moral standard is called **civil disobedience**. Civil disobedience requires nonviolently breaking the law publicly as an attempt to draw public attention to an injustice. Civil disobedience aims to bring about a change, often a change in the law that was broken. It requires a willingness to undergo whatever punishments the law provides for those who break that law. The term "civil disobedience" is applied even when the alleged injustice that a person protests is not an injustice of the law that the person violates but another law or legally sanctioned activity that one believes to be unjust. Recent examples include trespass on the grounds of nuclear weapons facilities and abortion clinics. Henry David Thoreau's 1849 essay on civil disobedience is a classic statement on the subject. [12]

It is also possible to change a law by a public protest that breaks no laws but attempts to use nonviolent means to draw attention to a perceived injustice, especially injustice in the law. **Nonviolent protest**, such as the Alabama bus boycotts that protested segregated busing, uses many of the same methods as civil disobedience but may not break any laws.

Conscientious refusal is a second related notion. Examples of conscientious refusal include refusal to carry out a work or military order that one believes to be immoral and refusal to eat higher animals or use products that have been tested on them. It can occur in work or nonwork situations and may or may not involve breaking any law. It may be done either simply from a motive of not participating in what one sees as a moral wrong or it may be done with the hope of making a public protest that will draw attention to the situation one believes is wrong.

Finally, there is **outright breaking and attempt to evade the law on grounds of conscience**. Refusing to turn over those whom the Nazis wished to exterminate (Jews, homosexuals, gypsies) is an example. Such action is covert, as contrasted with the publicity of civil disobedience, nonviolent protest, and conscientious refusal. It is morally justified only under conditions in which public protest would certainly be futile and a grave wrong is done if one complies with the law.

Sometimes all of the actions discussed in this section are loosely referred to as "civil disobedience"; however, if this is done, it is still important to distinguish between what are here called nonviolent protest, conscientious refusal, and a conscientious attempt to evade the law, since there are relevant ethical differences among these acts, especially in the conditions that justify their performance.

The purpose of this introduction has been to present basic ethical terms and distinctions are used in discussion of many ethical questions. The terms and distinctions are intelligible from a variety of cultural and religious perspectives, although some assume the context of a democracy.

Concepts with application to specific contexts of engineering practice and research will be introduced as needed in later chapters.

NOTES

1. These rough generalizations do not take account of differences among branches of these religions.
2. John Hick. 1993. *Disputed Questions in Theology and the Philosophy of Religion.* New Haven: Yale University Press, p. 93. For citations from Judaism, Christianity, Islam, Jainism, Confucianism, Hinduism, Buddhism, and African Traditional Religions, see http://www.silcom.com/~origin/ws/theme015.htm. This is part of a larger discussion of religious viewpoints on moral law that draws from a very broad group of religions in Part One, Chapter 2 of the World Scripture web site at http://www.silcom.com/~origin/wscon.html.
3. Amelie Oksenberg Rorty. 1995. "The Many Faces of Morality." *Midwest Studies in Philosophy*, XX:67–82.

4. These and other theories may be reviewed in a variety of sources. For example, see the articles under "Ethics" in the *Encyclopedia of Bioethics*, 2nd ed., New York: Macmillan, or for an on-line survey and bibliography see Lawrence Hinman's Ethical Updates at http://ethics.acusd.edu/index.html.

5. Ladd, John. 1979. "Legalism and Medical Ethics." In *Contemporary Issues in Biomedical Ethics*, ed. J. W. Davis, Barry Hoffmaster, and Sarah Shorten. Clifton, NJ: Human Press.

6. Personal communication, January 1996.

7. MacIntyre, Alasdair. 1984. "Does Applied Ethics Rest on a Mistake?" *The Monist* 67(4):499–512.

8. Robert Proctor. 1988. *Racial Hygiene: Medicine Under the Nazis*. Cambridge, MA: Harvard University Press.

9. Ladd, John. 1970. "Morality and the Ideal of Rationality in Formal Organizations," *The Monist* 54(4):488–516.

10. "Lotus – New Program Spurs Fears Privacy Could be Undermined." *The Wall Street Journal*, 13 Nov. 1990, p. B1 and "Lotus is Likely to Abandon Consumer-Data Project." *The Wall Street Journal*, 24 Jan. 1991, p. B1.

It is important to distinguish information that is private (rather than public) from information that is personal in the sense that it is information that would be intrusive for others to demand, obtain, or discuss. For example, in the United States in the early part of this century, some people (especially women) rarely disclosed their age. They considered this information highly personal even though their birth dates were matters of public record. The judgment of what matters are personal is highly cultural. For example, the Dutch consider it intrusive to look over the books in a person's bookshelf without first asking permission. In some cultures it is considered impolite to speak of a woman's pregnant condition even when it is evident.

11. Madison Powers. 1993. "The Right of Privacy Reconsidered." Commissioned paper for *Genetic Privacy Collaboration*, April 22, 1993, 8–22.

12. Henry David Thoreau, "Civil Disobedience." Reprinted in *Civil Disobedience in Focus*, pp. 28–48, edited by Hugo A. Bedau. New York and London: Routledge, 1991.

1

ETHICS AS DESIGN: DOING JUSTICE
TO ETHICAL PROBLEMS

Suppose I face an ethical problem; how ought I go about figuring out what to do? The question is not simply how should I evaluate proposed courses of action, but how should I go about devising such courses of action.

Ethical judgments are important in devising responses to ethical problems, of course. These judgments come in many forms, from "What is being proposed is morally wrong" to "This safety factor (or margin) is sufficient for the circumstances in which this device (or process or construction) will operate." This book is at least as concerned with devising good responses as with making ethical judgments.

People confronted with ethical problems must do more than simply make judgments; they must figure out what to do. (This is the reason for calling people "agents.")

Suppose my supervisor tells me to dispose of some regulated toxic substance by dumping it down the drain. In this case, part of my problem is that I have been ordered to

This is a subject on which, as Stuart Hampshire observed in 1949, ethics has had little to say. Hampshire made his point by saying that courses in ethics only teach students to critique moral actions rather than to resolve ethical problems. Writing *Innocence and Experience* some forty years later, he found the situation no better. [1]

As Hampshire pointed out, an agent (that is, the person who faces the problem) needs the skills of a judge in weighing alternative courses of action once these are formulated. But the skills of a judge are only part of the skills an agent needs to respond to an ethical problem. The rest of the task is a constructive or synthetic one of devising and refining candidate responses.

do something that is potentially injurious to human health and, furthermore, illegal. Assuming my supervisor knows, as I do, that the substance is a regulated toxic substance – an assumption I should verify – then my supervisor's order is unethical and illegal. This is an example of an ethical judgment that I make in describing the situation.

In this case the question is what can and should I do? It is not enough to say that I should not dump the waste down the drain. My problem is not the simple choice of answering yes or no to the question of whether I should follow the order. I need to figure out what to do about the supervisor's order. Shall I ignore it? Refuse it? Report it to someone? To someone else in the company? To the Environmental Protection Agency? Should I do something else altogether? Is there any place I can go for advice about my options in a situation like this? What are the likely consequences of using those channels (if they exist)? Where could I find out those consequences? Also, what do I do with that toxic waste, at least for the present? These are questions with important implications for human well-being, for fairness to others, and for the environment, as well as for my relationship with my supervisor and future with this company. Answering the question of what to do will depend on a variety of factors. Learning what factors to consider and how to assess them are components of responsible professional behavior.

The importance of finding good ways of acting (and not merely the ability to come up with the right answer to a whether question) may be brought home by an example. When did you or I last pour paint solvents, petroleum wastes, acetone (nail polish remover), motor oil, garden pesticides, or other household hazardous waste down the drain (or put spent batteries in the trash)? Was it before we knew it is a bad idea to dispose of these items in this way, or did we know it was a bad idea, but did not know what else to do with the refuse?

The need for a response is what makes ethical problems practical problems. The similarities between ethical problems and another class of practical problems, design problems, are instructive for thinking about the resolution of ethical problems and correcting some common fallacies about ethical problems.

Practical problems may or may not have solutions. Of those that are ethical problems, some call for coping rather than for solution. The perennial problem of human vulnerability, suffering, and mortality are such problems. Both ethical problems that call for solution and those that call for coping have their counterpart in design problems, although good ways of coping are also called "solutions" in the case of design problems. For example, design of a system of drainage ditches to prevent damage from periodic flooding of a nearby river counts as a solution to the problem of how to cope with periodic flooding, although the drainage ditches do not keep the river from flooding.

Design problems are problems of making (or repairing) things and processes to satisfy wants and needs. The analogy with ethical problems holds for a variety of design problems, from designing or repairing a bookshelf to devising a rotating work schedule, to designing or redesigning an experiment. The analogy between ethical problems and problems of engineering design proves to be especially instructive. Because engineering design is part of the university curriculum, much has been explicitly articulated about the design process in engineering. In contrast, craft skills are often transmitted by apprenticeship and articulated only in ways peculiar to a specific craft. Furthermore, engineering design stands out among college subjects in giving sustained attention to the synthetic reasoning necessary to construct good responses to practical problems. Because engineers recognize the importance of engineering design as well as engineering theory, they appreciate the importance of practical as well as theoretical problems and of synthetic as well as analytic reasoning. Devising a good response requires synthetic reasoning. Ethics has been more involved with analytic reasoning and the analysis of ethical problems and possible answers to them. Analysis is important, but not sufficient, to devise responses.

DESIGN PROBLEMS

Engineers recognize the ability to analyze the designs of others (i.e., being an astute judge of designs) as a useful skill for designers to possess, but not sufficient to make a person a good designer. For this reason, most engineering schools offer courses in engineering design that are markedly different from the engineering theory courses that teach students to apply theory to the solution of problems. The applications problems typically have unique, mathematically exact solutions.

The products of design may be single objects (e.g., a bridge at a given site) or a type of object (e.g., a new type of toaster) or process (e.g., a cost-effective way of making newsprint from recycled newspapers or a process for making weather-resistant paint). Design problems are familiar in engineering and science, but problems of engineering design (and experimental design) are especially instructive for present purposes. Not only is the design process well studied in engineering, but design problems in engineering are typically highly constrained, as are challenging ethical problems. The design process, especially in the ways in which it differs from merely analyzing the designs of others, highlights the very aspects of the agent's response to ethical problems that philosophy and applied ethics have had difficulty illuminating.

To develop a good response to an ethical problem, one must typically take account of a variety of considerations. In situations like the one just described, which raises questions of blame – for either negligence or intentional wrongdoing – fairness is a prominent consideration. Some tension or conflict may exist between the moral demands or values associated with these considerations, but it is often possible to at least partially satisfy many of these demands simultaneously. Indeed, doing so is a mark of wisdom. This seemingly common-sense observation about ethical problems has been obscured in recent years by a preoccupation with construing ethical problems as irresolvable conflicts between opposing principles or obligations. Although such conflicts are occasionally irresolvable, the initial assumption that a conflict is irresolvable is misguided, because it defeats any attempt to do what design engineers often do so well, namely, to satisfy potentially conflicting considerations simultaneously.

THE DESIGN ANALOGY

To illustrate the characteristics of a design problem, consider the design of a mechanically simple object: a child seat to fit on top of suitcases with wheels designed to be wheeled on board an airplane and stored under the seat or in the overhead bin. When removed from the suitcase, the seat must double as a child seat that will strap into an available airline seat, and the child seat itself must also fit easily into the overhead compartment. Several manufacturers make such suitcases. Most have similar features, making it possible to design a child seat that fits most of the suitcases in use. This is a design project in which I directed three mechanical engineering students. One student, Colleen, investigated what the potential user would require in such a device – such as ease of cleaning and having a place in the seat to carry a bottle, a pacifier, and similar paraphernalia. Two other students, Lisa and Kimberly, investigated standards and safety requirements and built rough prototypes. These demonstrated solutions to the design problem and developed some of the features of such solutions.

Lisa's and Kimberly's designs are significantly different solutions to the suitcase child seat design problem. For example, in Lisa's design, the horizontal crossbar that holds the child in place pivots around its permanent attachment to the end of the right armrest and its other end secures into the other armrest. In Kimberly's, the crossbar and armrest form a single U-shaped piece that lifts overhead like an old-fashioned highchair tray. (Both designs have the advantage that they do not detach from the rest of the chair, so they will not become lost.) Kimberly's design has larger dimensions. A larger seat might better suit a heavier child but would be more expensive to manufacture. Lisa's seat would accommodate most children under two years old, the age at which infants fly free with an adult.

These differences illustrate the first point about design problems that is significant for ethical problems: **For interesting or substantive engineering design problems, there is rarely, if ever, a uniquely correct solution or response, or indeed, any predetermined number of correct responses.** This is in contrast to puzzles, math problems, and most of the problems that engineering students typically do in problem sets.

People do speak as though "doing the right thing" is possible, but this way of speaking should not be taken to imply that ethical problems have unique correct solutions or responses. The view that ethical problems have unique correct solutions is more plausible if one starts from the assumption that possible responses to ethical problems are determined in advance and fairly evident. That would make ethical problems a type of multiple-choice problem.

There may be no solution to a given design problem – no way of making a thing that answers a certain set of specifications. (Perhaps no design of the child seat would make it both light enough to be supported by a soft-sided suitcase and strong enough to meet safety requirements, for example.) However, if one solution to a design problem exists, others usually do as well.

Some engineering design and ethical problems may be trivial in that the specification of the problem leaves little leeway in an acceptable solution. The question of what to do about a promise that one has freely made, in circumstances where no morally compelling counterclaims exist, is trivial in this sense: One should keep the promise. So is the design of a bolt to fasten the housing of the radar for a large commercial aircraft. Only a few questions would be open, such as whether the bolt should be made of corrosion-resistant material and whether this bolt should be interchangeable with many other bolts used in the aircraft. In both the promise-keeping and bolt-design cases, devising an appropriate response is not demanding, so the principal moral question is whether one is sufficiently conscientious in acting to accomplish the goal.

It is for nontrivial ethical problems that the analogy with problems of engineering design is most important. The fulfillment of moral responsibilities in general, and professional responsibilities in particular, provide many examples of ethical problems that resemble interesting design problems. If ethical problems are not assumed to take the form of multiple-choice problems, it is not surprising that where there is one course of action that provides an ethically responsible resolution of an ethical problem, other responses may also be acceptable.

The initial problem about the toxic waste constitutes an interesting ethical problem with several acceptable responses. It may be possible to change the supervisor's mind, perhaps by detailing the potential health effects or the legal liability to the company, or by simply stating that I cannot in conscience dump the waste. If the supervisor is adamant, it may be possible to get others in the company (the ethics or environmental office, if any; the legal department, if any; etc.) to countermand his order. The character of my organization makes a difference to my response, too. Although some organizations have a strict chain of command, others, including most universities, make a point of having "multiple channels" for working through many problems. There may be several ways of properly disposing of the waste while not embarrassing the company or coworkers more than necessary.

This brings me to the second point about design problems: **Although no unique correct solution may exist, nonetheless, some possible responses are clearly unacceptable – there are wrong answers even if there is not a unique right answer – and some solutions are better than others.**

One commonly hears the assertion that for some or all ethical problems, "there are no right and wrong answers." Those who say this may be attempting to acknowledge that there are no unique correct solutions to ethical problems, or they may be espousing an extreme relativism in ethics. Some possible responses to moral and design problems are so poor as to be clearly wrong. "Intimidate vulnerable parties into acquiescing to what we want" or "make it with a safety factor of 1" (which means no safety margin) are wrong answers. The first violates basic moral standards; the second violates basic safety standards. A child seat that could not recline when in the airline seat would be more of an irritation than a comfort to the child and accompanying adult. A suitcase child seat lacking any safeguard would be prohibitively dangerous should the handle slip out of the adult's hand. In our problem of being ordered to dump toxic waste, dumping the waste down the drain is a wrong answer, as is dumping the waste under the supervisor's hedge. (As these examples illustrate, some design questions are also questions of ethical responsibility.)

This leads me to a refinement on the first point: **Although for interesting or substantive engineering design problems there is rarely, if ever,**

a unique correct solution, two solutions may each have advantages of different sorts, so it is not necessarily true that, for any two candidate solutions, one must be incontrovertibly better than the other.

In the case of Lisa's and Kimberly's designs, one is not clearly better than the other, although some features of one are clearly better than the corresponding features of the other. If no design feature were constrained by the design of some other feature, it might be possible to collect together the best features into one best design. However, some features are so constrained. For example, the design of the security strap that fits between the child's legs and runs between the crossbar and the seat depends on the design of the crossbar. Furthermore, even a given feature may be better in some respects (easier to keep clean, more comfortable for the youngster, less expensive to manufacture) and worse in others (more cumbersome for the adult to operate, more likely to break). Such a feature may be advantageous for some users and disadvantageous for others.

For ethical problems, too, different courses of action that all satisfy basic constraints may have different advantages. Suppose, for example, in the case of the disposal of the toxic waste, my supervisor is acting on habits established in the 1970s and 1980s, when such dumping of waste was far more prevalent. It is possible that my supervisor may not appreciate what is wrong with dumping, or not know he is violating the law in doing so. Significant changes in knowledge, regulation, and company attitudes have taken place in the past two decades. My supervisor may then be open to arguments that things have changed, especially if they come from his boss or from our environmental or legal department. Before I go to those sources, I might tell him that I think we should get their view and that I intend to do so. That forewarning may prevent his feeling undercut when I take the concern further, and it may even convince him to do so.[2]

Suppose, instead, I take the approach of saying that I cannot in good conscience dump the waste. This has the advantage that it does not raise the specter of my continually going over his head (an effect that may be more of a danger if I am new to the job). It also, however, leaves less opportunity for him to find out that standards have changed. Suppose that in response to my conscientious objection he says, "Well, if you are so squeamish, I will do it myself." What, if anything, do I say then? What if he then proceeds to dump the waste? Each of the two avenues that I have outlined has its own advantages and disadvantages. It may not be possible for me find out ahead of time which one is more likely to work well with my particular supervisor.

Notice that many of my subsidiary judgments are about how far to go to convince my supervisor of the error of his or her order. I must first think of what further actions I might take before considering whether to take those actions. This point is worth emphasizing because, as I mentioned, ethical problems have often been represented as choices of which of two (or more) options to take, that is, as decision problems between prescribed alternatives and thus implicitly as multiple-choice problems. Most common ethical disagreements are about how far to go in trying to achieve some end that both sides agree is at least somewhat desirable or in avoiding violating some other ethical norm that both sides agree upon. Relatively few arguments have the form of one side thinking that some ethical value is a noble one and the other side thinking it is pernicious or of absolutely no importance.

Nothing in the argument for a multiplicity of acceptable ethical responses requires ethical relativism. The variety of acceptable solutions to complex ethical problems does not require agents holding a variety of moral beliefs. Even if all agents have exactly the same moral beliefs, different responsible actions frequently may be found. Advantages and disadvantages of acceptable solutions may differently suit the life circumstances of different individuals independent of their moral beliefs.

A third and final point about solutions to design problems that holds for responses to ethical problems as well is that they must do all the following:

- **Achieve the desired performance or end, e.g., create a child seat that fits on a wheel-on-board-suitcase or fulfills one's responsibility for environmental safety.**
- **Conform to specifications or explicit criteria for this act; for example, the seat must fit inside the overhead rack and be a comfortable booster seat that straps into an airline seat; straightening out the toxic waste issue should not take so much time that you fail in other major responsibilities.**
- **Be reasonably secure against accidents and other miscarriages that might have severe negative consequences.**
- **Be consistent with existing background constraints; for example, for the child seat: do not require very expensive, scarce, or hazardous materials to manufacture; for any ethical problem: background constraints include the requirement to avoid violating anyone's human rights (so it goes without saying that even if feasible, killing off the supervisor is not an option in response to the order to dump the toxic waste).**

FOUR MORAL LESSONS FROM DESIGN PROBLEMS

The analogy between ethical and design problems suggests some strategies for addressing ethical problems.

Consider the examination of the situation and definition of the problem. Some assessment is needed just to name the problem. In the case of design problems, the ambiguity is typically limited by lack of knowledge of what potential users might require in such a device (and hence the constellation of features in such a device) and of what solutions are already available. Often it is not clear how far you can go in meeting some requirements and still satisfy others. For example, in the case of a child seat it would be desirable to accommodate large toddlers and three-year olds, as well as average-size two-year olds, but such suitcases will support only a limited load.

Another design problem illustrates the need at early stages to take actions that take account of ambiguity or uncertainty. Consider a device that automates testing for a variety of immune factors. This complex device was so novel that when it was designed, there were no industry standards for the characteristics of such a device. At the initial stage, designers had to decide such questions as how constant to maintain the temperature at which the device maintains chemical reactions: Should the specifications be for a temperature of $37°C + 1°$ or $37°C + .1°$? Once such specifications were decided upon, the designers built a feasibility model, that is, a model that meets the specifications and embodies the core features of the technology. Such a feasibility model demonstrates that it is possible to create the device in question but typically leaves open many questions about the device that will be manufactured and sold. (In the case of a device as complex as this one, "engineering models" are then built. These include some user interfaces (that is, some of the controls that a lab technician running the assays would use), which are similar in fit, form, and function to the device that will eventually be manufactured and sold. After the engineering model comes an "engineering prototype," which is an economically and technically manufacturable version of the device. Next, "manufacturing models" are built on the manufacturing floor to detailed documentation to catch any problems that arise when the device is actually manufactured. Finally come production units that can be sold.)

The initial phase of the design of the immuno-assay device illustrates the task of problem definition. Engineers recognize the importance of allowing for as much flexibility as possible in the definition of the problem, that is, to avoid foreclosing options to change features or to add new ones in successive models to improve safety, performance, reliability, or manufacturability. Comprehensive foresight prevents difficult or costly changes when far along in the process. For example, retooling for manufacture (changing the manufacturing process) is very expensive.

The first lesson from design problems for ethical problems is to begin by considering the unknowns and uncertainties in the situation. In the case of ethical problems, the situation may even be fundamentally ambiguous, creating even more of a challenge for foresight. At least with a design problem it never turns out that what seemed to be a problem of designing an airplane turns out to be a problem of designing a coffee pot. In contrast, if one hears from one person that another is doing something wrong, it may be that the second is doing wrong or that the first is slandering the second. The only thing certain at the beginning is that something is not as it should be.

Appreciating ambiguities and uncertainties is important. These are often underemphasized in professional ethics. For example, the original (1989) edition of *On Being a Scientist* (the handbook on research and research ethics for young scientists put out by the National Academy of Sciences, the National Academy of Engineering, and the Institute of Medicine) recommends that when one believes one has witnessed research misconduct, one should talk it over with a trusted experienced colleague and "[o]nce sure of the facts, the person suspected of misconduct should be contacted privately and given a chance to explain or rectify the situation." Two things are wrong with this piece of advice. First, as is now widely recognized, confronting a person who has committed research misconduct frequently leads to data destruction or other attempts at concealment. Therefore, in the face of strong evidence that misconduct has occurred, and having talked over the matter with a knowledgeable and trusted person or institutional ombudsperson, the matter may best be turned over to one's institutional research standards officer. However, the more general point is that it is often not possible to wait for certainty before responding. The advice to act only when one is sure of the facts is advice to avoid action.

What is needed are ways of acting that will prove prudent and fair however uncertainties are resolved. In cases in which crucial ambiguities cannot be fully resolved early in the situation, the ambiguity should be understood as a defining characteristic of the situation – for example, ambiguously either malfeasance or slander. Faced with an ambiguous problem, agents typically need to figure out whether to gather more evidence, how to raise the issue (or gather more evidence) without being unfair to others, and how to best elicit support for their concern to achieve a fair resolution. When reporting an ambiguous situation to others, one good ethical rule of thumb is to clearly state the facts as one has them with a minimum of interpretation.

The second lesson from engineering design for ethical problems is that the development of possible solutions is separate from definition of the problem and may require more information. This is one of many features that distinguish ethical problems from formal "decision problems." "Decision problems" or problems in decision analysis include specification of the alternatives among which one is to decide. Therefore, a fully defined decision problem is a type of multiple-choice problem. The need to *develop* possible solutions in real life shows that open-ended statements of ethical problems do more justice to them than do representations of them as multiple choices.

Furthermore, before proposing solutions, agents must frequently clarify the problem. Although open-ended statements do more justice to ethical problems than do multiple-choice statements, even open-ended statements are only outlines of ethical problems. If one had an actual ethical problem, there would be real details to be examined. For example, if I really had the problem about the disposal of toxic waste, there would be a particular person who would be my supervisor whose character I might learn more about. There would be an actual organization (a company, a university) with particular policies that I could investigate. One of the important characteristics of a responsible or wise response to a practical problem is appropriate investigation of a problem before attempting to solve it. This is certainly something that engineers do with design problems.

Part of this investigation for design engineers was already mentioned: investigation of the requirements of potential users. This is especially important for an entirely new device. If engineers are seeking to design a better mousetrap, say, then they engage in "benchmarking," that is, they gather information about the mousetraps already available. Just as important, they do an investigation of the demand for features not currently available in mousetraps and of the relative importance of features in the mind of the user. In U.S. engineering the demands of the user are often called "the voice of the customer"; however, the user is not always the buyer. In Sweden, for example, the workers who will use a new medical device are consulted on its design.

Too often when statements of ethical problems are presented, students' attempts to interrogate the problem are cut off. Answering problems without seeking to investigate them is poor preparation for understanding and addressing actual ethical problems.

From the place of brainstorming in the practice of engineering design, we learn more about how an agent goes about developing responses. Brainstorming requires an uncritical atmosphere in which people can present "half-baked" ideas that may be later refined or combined. Articulation of any half-baked ideas is discouraged in the many ethics classes where adversarial debate comprises the primary method used. Although adversarial debate format classes may provide some useful pre-law training, it does not help develop the ability to think constructively about resolving ethical problems.

A rather heroic capacity to brainstorm in the face of critical response is demonstrated in the responses of a child, known as "Amy," who was asked to respond to the "Heinz dilemma." When Amy is asked if Heinz should steal a drug he cannot afford to save the life of his wife, she proposes new alternatives to either stealing or letting Heinz's wife die:

> Well, I don't think so. I think there might be other ways besides stealing it, like if he could borrow the money or make a loan or something, but he really shouldn't steal the drug – but his wife shouldn't die either.

Asked why he should not steal the drug she replies:

> If he stole the drug, he might save his wife then, but if he did, he might have to go to jail, and then his wife might get sicker again, and he couldn't get more of the drug, and it might not be good. So, they should really just talk it out and find some other way to make the money.

> – Gilligan, 1982, pp. 27, 28

The brainstorming (in this case "borrowing the money" or, as Amy elsewhere suggests, persuading the druggist to lower the price) and interrogation of the problem are not entirely separable activities. In addressing design problems, suggestions from potential users about their needs frequently stimulates new ideas, and ideas for approaches to the design may stimulate new questions for potential users. For ethical problems, additional information gained through interrogating the problem frequently changes the desirability of possible responses.

The point is illustrated by consideration of the following situation: A highway safety engineer is allocating resources for safety improvements and considers two intersections. Both have the same number of fatal accidents per year. However, one is in a rural setting and the other is in an urban setting. The urban intersection handles on average four times the number of cars as the rural intersection and also has a higher rate of minor injuries and property damage than does the rural intersection. There is just enough money in the budget to improve one intersection. Which one should it be?

The choice of improving the urban intersection is often justified on the ground that, improvements there will have the greatest overall reduction of injury, and this choice is cited as illustrating a utilitarian choice of "the greatest good for the greatest number." The choice of the rural intersection is made on the ground that it is a more dangerous intersection in the sense that the likelihood of a fatal accident for a given use of the intersection is four times higher. This consideration is taken to represent concern for fairness (presumably equal distribution of the risk of fatal injury associated with going through any given intersection) or even respect for individual rights.

What is relevant here is not how well this story illustrates the philosophical distinctions between utilitarian and competing foundationalist schools of thought in ethics, but the danger that this example will be misunderstood as an example of problem solving. Notice first that the problem is presented as a forced choice between spending all the remaining resources on one intersection and spending it all on the other. In fact, there would likely be many other choices. For example, putting up traffic signs at both intersections may be an alternative to installing traffic lights at either one. However, even accepting the multiple-choice character of the problem as stated, there is a great deal of potentially relevant information that the example does not tell us about the accidents. For example, suppose that at one intersection, but not the other, in all serious accidents at least one of the drivers involved was drunk (or fell asleep or had a heart attack, etc.). Such information might show that the most crucial variable for reducing serious accidents at one site is reducing driver impairment, while at the other it is the physical characteristics of the intersection; the latter would be best remedied by changing the intersection itself.

A third lesson from design problems concerns acting under time pressure. It is often important to begin by pursuing several possible solutions simultaneously, so that one will not be at a loss if one meets insuperable

obstacles, but still avoid spreading one's energies too broadly. This admonition applies both to the design of individual features of the device and to approaches to revising the design when obstacles are encountered at later stages. For the immuno-assay device, the possible solutions were of three general kinds: mechanical modification, modification of the chemical procedure, and modification of the software. Since modification of the software was generally the cheapest modification, that was preferable where it could provide the requisite fix.

The need to act under pressure of time is also a common feature of ethical problems. In the face of time pressure it is reasonable to pursue several possibilities simultaneously in case one fails to prove practicable. Consider the ideas proposed by Amy, the child who rejects the forced choice of the Heinz dilemma and "brainstorms" a variety of possible courses of action. (Some of these have to do with relationship, such as remonstrating with the druggist; others, such as "taking out a loan," do not.) The simultaneous pursuit of several options is a mark of good design strategy when the danger exists that one line of development may prove unfeasible. Pursuing several options contrasts with representing the ethical problem as a static situation with static solutions; the problem becomes simply one of selecting the right alternative and doggedly pursuing it.

Fourth and finally, the dynamic character of problem situations has further implications. Both the problem situation and one's understanding of it are likely to change and develop over the course of time.

For example, in attempting to avert the *Challenger* accident, engineer Roger Boisjoly's problem situation began with evidence, in the form of blackened grease, that hot gas was escaping through the joints. The problem then became one of conducting experiments to test the effect of temperature on the seals, and then one of getting a team formed to redesign the seals and of getting resources to do so. The final problem became one of stopping the flight in view of predicted record cold temperature.

THE DYNAMIC CHARACTER OF ETHICAL PROBLEMS

When the dynamic character of the ethical situation is neglected, people often mistake doing the wrong thing and then making the best of the bad situation with taking an action that is justified in some circumstance. For example, a colleague with whom I was working to formulate some criteria for research ethics raised the question of whether gift authorship is ever ethically justified. (Gift authorship in a research context is the listing of a person as an author although the person has not contributed substantially to the research reported in the paper. Issues of fair credit and authorship will be discussed in Chapter 9.) My colleague was thinking of a case in which he had an idea for a collaborative project and proposed it to a European researcher who had done some work on a subject that established some of the groundwork for the new effort. The European researcher at first expressed interest but then failed to respond when my colleague actually proposed to start the work. After several communications brought no response, my colleague undertook the work with members of his own lab only. There was some delay because my colleague's group had to recreate some research materials that would have been on hand at the European researcher's lab. In due course my colleague and one of his post-doctoral fellows completed the research and wrote a manuscript reporting the work. As a courtesy, because the work was built in part upon the earlier work of the European researcher, my colleague sent a "pre-print," that is, a copy of the unpublished manuscript, to the European researcher. That researcher replied that my colleague could not publish the paper because the virus he had used had been obtained from the European researcher's lab for a different purpose and the European researcher had not given permission for the new use. (The sharing of research materials, or the means for making them, is encouraged in science, although no lab is expected to take on great burdens to supply others with materials. Some science and engineering journals require that those who publish articles in their journal furnish to others the reagents and similar materials necessary to replicate the work. In this case, however, my colleague had agreed to use the virus only for a single purpose. That purpose did not include making materials for the project described in the manuscript.)

Hoping to shame the European researcher into desisting from his complaints, my colleague wrote back asking if there was someone from his team whom the researcher thought should be added as an author on the manuscript. To my colleague's dismay, his European colleague sent a letter back nominating both himself and a post-doc in his lab as coauthors. My colleague's post-doc was about to take a job in proximity to the European colleague. Because this post-doc, who had done nothing wrong, would be vulnerable to retaliation, my colleague decided to go ahead with gift authorship. He added the names of the two European researchers to the list of authors on the manuscript. He did this despite his firm conviction that gift authorship is a corrupt practice.

Many would agree that my colleague "made the best of a bad situation." His story was not one in which gift authorship is *justified*, however, because the situation itself was one that was partly his own doing and one that, as an ethical matter, he ought not to have gotten himself into. His is a cautionary tale. Cautionary tales help others avoid the same pitfalls. In the future he would take care not to get into this situation; he will be more careful to check the conditions under which he receives research materials and would not again make the mistake of offering gift authorship as a backhanded form of moral criticism. (It is hard to imagine a similar situation that one could get into without making such mistakes.) Everyone makes mistakes, but responsible people exercise care not to make many and learn from their mistakes. Wise people learn from the mistakes of others and so do not have to learn everything "the hard way."

In this book we will often be concerned with learning how to anticipate the possible consequences of actions to avoid ethical pitfalls that place a person in a position of having to choose the lesser evil.

PROBLEMS AS EXPERIENCED BY AGENTS

Here are two problem situations that illustrate some of the points we have been discussing, especially coping with ambiguity in responding to an ethical problem.

SCENARIO: WHAT ABOUT MY CONTRIBUTION?

You are a third-year graduate student. You started graduate research with Prof. One in Great Lab working on the Fantastic project. By the end of the first year you had not only become proficient at many of the more routine tasks of the project, but you had made one small but notable refinement to the approach to the segment assigned to you. At the end of the first year, Prof. One went on leave for a semester and you started working with Prof. Two in the same lab but on a different project. Prof. One returned for the spring semester and took up the Fantastic project, among others. The following fall, the beginning of your third year, you learned from another student who was working on Fantastic that Prof. One is publishing a paper on some aspects of Fantastic with this student, a paper that contained your refinement.

What, if anything, can and should you do?

Are there any ambiguities in the situation? If so, how can you fashion a response that will be appropriate however the ambiguities are resolved? (Remember what was said above about ambiguities in ethical problems frequently being greater than the unknowns in a conceptual design problem.)

You do not know whether Prof. One remembers your involvement and contribution; remembers it but judges your contribution as insignificant; judges it significant but worthy of an acknowledgment only, and not joint authorship; or perhaps even is planning to add you as a third author. In resolving the situation, it will be important to attend to alternative possibilities.

Because there is often the possibility that the situation is not what it first seems and that a possibility may have escaped your attention, the previously stated rule of thumb – to clearly state the facts as one has them with a minimum of interpretation – has special ethical significance when there is a question of another's negligence or malfeasance. It is ethically important to be fair to others and avoid spreading rumors about them. Minimizing your interpretation will minimize the possibility that you have interpreted the situation the wrong way. (The same rule is also simply *prudent*, in that sticking to the facts makes it less likely that raising the issue will get you into trouble.) As with all rules of thumb, the advice does not always apply. If a larger overall pattern seems apparent – for example, the person whose actions are in question has a pattern of cheating in some way or harassing people – then the best idea may be to raise the possibility of that interpretation, stating the facts that lead to that interpretation. However, the question of that overall pattern is often best raised with some designated neutral (like an ombudsman in an organization) or a person who has official responsibility and some experience in looking into such matters.

SCENARIO: IS IT PLAGIARISM?

You find that two academic publications have remarkably similar text expressing an idea that is not part of common knowledge in your field.

What if anything can/should you do, and how ought you go about it?

The similarity between the texts in this scenario requires explanation. There may be intentional wrongdoing, namely plagiarism (the first author of the second or the second author of the first), but this is a matter that cannot be confidently decided from publication dates alone. The plagiarist may have seen the work of the plagiarized but published first. Furthermore, especially if the idea came up in conversation, the person who received the idea may no longer remember where he first heard it, or even that he did hear it from another person, so the fault may be one of negligent or reckless misappropriation rather than plagiarism.

The situation may be one that is very different from plagiarism, however. Perhaps the ideas were original to a third party who was the teacher of both and who gave each the mistaken impression that those views were common knowledge in the field. Therefore, in this case too, it is important for the person who raises the issue to do so in a way that does not prejudge the issue.

MAKING AND ASSESSING ETHICAL JUDGMENTS

Understanding, assessing, and making ethical judgments compose a significant part of learning how to respond well to ethical problems. Considering the judgments of experienced and authoritative individuals and bodies is a good way of seeing what factors these experienced individuals find relevant. It is also helpful in discovering the characteristic priorities of specific individuals and groups whose opinions are influential in the sphere in which one is working. In this book I often refer to or summarize cases, decisions, and ethical opinions issued by the National Society of Professional Engineers' (NSPE's) Board of Ethical Review (BER), as well as referring to ethical guidelines of other ethically active professional societies like the American Chemical Society (ACS), the National Academy of Sciences, and the U.S. Supreme Court; occasionally I use opinions from ethics offices of companies that employ large numbers of engineers. The reasoning offered in these judgments and the factors recognized as morally relevant exemplify the feature of some complex moral judgments on some relevant situations. Judgments by these particular bodies do, of course, reflect the particular mandate of the organizations they represent. Ensuring that laws are consistent with the U.S. Constitution is of paramount concern to the Supreme Court but may not be for other bodies making moral judgments. The NSPE is concerned with preserving the cohesiveness of the engineering profession. Therefore, as they affirm an engineer's responsibilities and obligations concerning public welfare, clients, and employers, the NSPE asks engineers to go to considerable lengths to show consideration to fellow engineers in fulfilling these responsibilities and obligations. Not surprisingly, while setting expectations for engineers to fulfill their responsibilities for the public and others, the ethics offices of reputable companies encourage their engineers to do so in ways that will minimize the likelihood of damage to the company. Therefore, these judgments are not intended to be the uniquely correct assessments of the cases on which they are offered.

Below is a problem situation that lies outside the domain of situations that the NSPE BER, the U.S. Supreme Court, or corporate ethics offices would normally consider, although many of the ethical values that those bodies affirm are relevant to it. Unlike the cases presented by the NSPE, this scenario, like most in this book, has some realistic uncertainties. As we saw earlier, these uncertainties may make the nature of the ethical problem ambiguous.

SCENARIO: RISKY RACING[3]

I am a new member on my university's solar car team. In just two months a prestigious race will take place in Australia and we're frantically trying to get our car to run properly. Sixty entries from all around the world will compete for the glory, prestige, and sponsorship a good performance brings. Media coverage will be extensive.

One of the main selling points we used to attract our sponsors was how lightweight and small our car is in comparison to others. Now it appears that the lightness of our car may be a liability. Last week during a test run, the car spun out of control after a moderate 45 mph turn and came to rest 100 feet into someone's front lawn. The driver wasn't injured, but was so shook up that she refused to ever drive the car again. Just two days later, the car slid off the road and flipped over after passing through some puddles left from a previous day's rain. The new driver got away relatively unscathed but probably would have been killed had the car veered into the other lane of oncoming traffic.

Some of the team members quietly admit that the car is an overpowered 3-wheel torpedo. But much time and money (over $40,000) has already been invested in the project. The plane tickets to Australia and the entry fee for the race are nonrefundable. The race is only held once every three years, and the team didn't enter the last one.

To a large majority of the team, this is the only big race they'll ever compete in.

The team leadership reasons that since the roads in Australia are likely to be long, flat stretches, the car should probably make it through the race without a serious accident. I am not convinced. The car will have to travel at 60 mph for most of the race to be competitive. Very large 80-wheeled trucks known as "land trains" commonly barrel down the race route at 70 mph. Because I'm a new member and I have not put as much time into the project as some of the others, my opinions don't hold much weight.

What can and should I do?

SUMMARY AND CONCLUSION: IMPROVING ON EXCELLENCE

Part of the explanation for the misunderstanding and misrepresentation of ethical problems is that most of recent ethics and applied ethics have neglected the perspective of the moral agent. Instead, ethics has exclusively emphasized the perspective of the judge or that of a disengaged critic who views the problem from "nowhere" and treats it as a "math problem with human beings." For the agent facing an ethical problem, not only are possible responses undefined, but the nature of the problem situation itself is often ambiguous. As a result, the agent faces a whole series of smaller problems about what to do next in the face of multiple ambiguities and uncertainties. To be responsible requires consideration of how to treat others and what becomes of others and oneself in addressing intermediary problems, as well as in the final outcome of the larger story within which the smaller problems reside.

The understanding of the design activity in engineering, especially in the ways in which it differs from merely analyzing existing designs, highlights the aspects of the agent's response to ethical problems that philosophy and applied ethics have had difficulty illuminating. The *multiply constrained* nature of many problems in engineering design provides an excellent model of challenging ethical problems involving many types of moral considerations, all of which must be taken into account. Many ethical problems that are represented as conflicts are better understood as problems with multiple demands and ethical constraints, constraints that may or may not turn out to be simultaneously satisfiable.

The analogy with design problems implies that we should expect that even excellent responses to a problem may be improved upon in many cases. I embrace this implication. To frame ethical problems exclusively from the vantage point of the judge or the moral critic, to the neglect of that of someone facing the problem, associates ethics with judgment and criticism and creates incentives for people to insulate themselves from criticism, either by narrowing the scope of the problems they address or by developing ready rationalizations for their behavior. However, pressing problems, both individual problems of how to be a good engineer, teacher, parent, or friend and social problems such as the provision of good health care or protecting the environment, are multiply constrained problems that require continuing input and oversight by many individuals and organizations. Recognizing that good resolutions of ethical problems can be improved upon should have the salutary effect of promoting open, constructive, and nondefensive discussion of ethical problems.

NOTES

1. Hampshire, Stuart. 1949. "Fallacies in Moral Philosophy." *Mind* 58:466–482. Reprinted in *Revisions: Changing Perspectives in Moral Philosophy*, edited by Stanley Hauerwas and Alasdair MacIntrye, 1983, and Stuart Hampshire, 1989. *Innocence and Experience.* Cambridge, MA: Harvard University Press.
2. An opinion by the NSPE's Board of Ethical Review on a case that concerns going "over the head" of one's supervisor when one believes the supervisor to be in the wrong is Case No. 82-7.[www] It was originally published in *Opinions of the Board of Ethical Review Volume VI*, Alexandria Virginia: National Society of Professional Engineers, 1989, pp. 27–29.
3. Based on a scenario by Mike Wittig, MIT '95.

2

THE BASIS AND SCOPE OF
PROFESSIONAL RESPONSIBILITY

PROFESSIONS AND NORMS OF PROFESSIONAL CONDUCT

Professions are those occupations that both require advanced study and mastery of a specialized body of knowledge and undertake to promote, ensure, or safeguard some matter that significantly affects others' well-being. This chapter will examine the norms and standards of responsible conduct in professional practice. Ethical (and sometimes legal) requirements also exist on the practices of nonprofessionals whose work immediately affects the public good, of course. For example, food handlers are bound by sanitary rules. Furthermore, many moral rules apply equally in all work contexts. All workers have an ethical obligation not to deceive their clients or customers, for example. What is distinctive about the ethical demands professions make on their practitioners is the combination of the responsibility for some aspect of others' well-being and complexity of the knowledge and information that they must integrate in acting to promote that well-being.

Moral rules, such as the one against deception, are important, but professional responsibilities cannot be captured in such rules. Fulfilling professional responsibilities requires more than rule following. Fulfilling a responsibility requires some maturity of judgment. The expressions "the age of responsibility" or "the age of discretion" acknowledge the maturity of judgment required to take on responsibilities. Carrying out a responsibility requires making complex judgments that integrate a variety of considerations in deciding how best to achieve certain *ends*, such as safety. In general, fulfilling a responsibility is not reducible to completing a checklist of acts that one is morally required to perform (and refraining from performing those that are forbidden). Responsibility requires judgment.

74

Professional responsibilities are those that require acquisition of the special knowledge that characterizes a particular profession and application of that knowledge to achieve certain ends. The ends or results that the professionals in engineering, research, medicine, or law work to achieve or include are, respectively, worker or public safety, sound research results, a good health outcome for one's patient, and a good legal outcome for one's client. The professional must figure out in each case what acts will achieve the desired ends, and this requires complex problem-solving skills. The fulfillment of moral responsibility calls for skills analogous to those of a designer, as we saw in the first chapter. Because the members of a profession are entrusted with a matter that significantly influences human well-being, these members are expected to live up to certain ethical norms in their professional practice.

Although the requirement that an occupation have significant influence on human well-being preserves part of what was ethically significant in traditional designations of the learned professions, some writers on professionalism take "professional" to require high social status. So sociologist Bernard Barber, who has written extensively about trust in the professions,[1] counts scientists as professionals but classifies engineering and nursing as "quasi-professions," because most are employees.

Some scientific societies have only recently formulated a code of ethics. The American Physical Society issued their first ethics statement only in 1992, and that code deals only with matters of research ethics.

The professions that were first identified (for example, medicine and law) clearly addressed some aspect of human well-being. However, today many occupations, including, for example, florists, claim to be professions and support their claim by having a code of ethics. (The codes of ethics of occupations that do not fulfill the requirement of directly influencing a major aspect of human well-being may, nonetheless, have ethical content. Where they do, they commonly articulate norms or standards of fair business practice.)

A distinction is often made between a scientific, scholarly, "learned," or disciplinary society, on the one hand, and a professional society, on the other. The former focus on technical or scholarly advances in a discipline. The American Philosophical Association, for example, is a disciplinary or learned society. A purely disciplinary society has no code of ethics.

Societies like the IEEE have characteristics of both a disciplinary society and a professional society.

Professional societies such as the National Society of Professional Engineers (NSPE), the American Medical Association (AMA), and the American Bar Association (ABA) issue explicit statements in their codes of conduct. Although these statements have names like "code of ethics," as we shall see they vary in the extent to which the matters they address are matters of ethics.

HOW ETHICAL STANDARDS VARY WITH PROFESSION

In professional codes of ethics, rules of practice specify the acts that must or must not be performed. Here are such items from codes of professional behavior of several engineering societies:

> Engineers shall not disclose, without consent, confidential information concerning the business affairs or technical processes of any present or former client or employer, or public body on which they serve.
>
> – National Society of Professional Engineers (NSPE), *Code of Ethics for Engineers*[2]

> An engineer shall not accept a mandate which entails or may entail the disclosure or use of confidential information or documents obtained from another client without the latter's consent.
>
> – Ordre des ingenieurs du Quebec (OIQ),[3] *Code of Ethics of Engineers*

> We, the members of the IEEE, . . . do hereby commit ourselves to the highest ethical and professional conduct and agree . . . [10 items, including] to reject bribery in all its forms.
>
> – Institute of Electrical and Electronics Engineers (IEEE), *Code of Ethics*

> Engineers shall not solicit or accept financial or other valuable consideration, directly or indirectly, from contractors, their agents, or other parties in connection with work for employers or clients for which they are responsible.
>
> – NSPE, *Code of Ethics for Engineers*

> [Members] shall neither pay nor offer directly or indirectly inducements to secure work.
>
> – *Code of Ethics of the Institution of Engineers*, Australia (IEA)

Some rules of practice in one profession have counterparts in other professions; others do not. For example, rules about maintaining client confidentiality are found in law and health care as well as engineering.

A lawyer shall not reveal information relating to the representation of a client unless the client consents after consultation, except for disclosures that are impliedly [sic] authorized in order to carry out representation ...

– American Bar Association (ABA), *Model Rules of Professional Conduct*

The information disclosed to a physician during the course of the relationship between physician and patient is confidential to the utmost degree. ...The physician should not reveal confidential communications or information without the express consent of the patient, unless required to do so by law.

– American Medical Association (AMA), *Principles of Medical Ethics*

Unlike engineering codes, the codes for law and medicine have no rule against paying or accepting bribes, although other improper payment are forbidden. Of course, it is no more ethically acceptable for physicians than for engineers to accept bribes. However, being offered or being tempted to offer others an outright bribe is not a common moral hazard for physicians. As this example illustrates, codes are not intended as exhaustive specifications of what one should or should not do, but as guidance, especially for the new member of a profession, on handling common temptations and avoiding common pitfalls in handling problems that arise in the profession but may be unfamiliar in other contexts. [4]

Payments that could create a conflict of interest, and in particular kickbacks for referrals, are of concern for physicians' organizations. Among physicians the practice is called "fee-splitting." Fee-splitting is prohibited in the AMA code. A somewhat more remote analog to bribery is the lavish entertainment of physicians by drug companies. For example, until a few years ago many drug companies commonly invited physicians to all-expense-paid "educational" seminars in resort locations. Accepting such trips has been widely criticized, because at a minimum it creates the appearance that the drug companies are buying the favor of physicians. It has been significantly curtailed in the 1990s.

Such lavish entertainment has long been unacceptable by the standards of the engineering profession. The NSPE has stringent limits on acceptable levels of hospitality offered by vendors to prevent not only bribery but any conflict of interest or the potential for such a conflict.

To be in a **conflict of interest** position, a person must be in a position of trust, which requires him or her to exercise judgment on behalf of others (people, institutions, etc.) and have interests or obligations of the sort that might interfere with the exercise of sound judgment in that position of trust. Furthermore, the conflict of the interests or obligations deriving from the position of trust with the others must be widely acknowledged as the sort that are not well controlled by other means. (Such acknowledgment changes over time.) As a result, the person is morally required to either avoid the situation entirely or, if that is impossible or excessively burdensome, to openly acknowledge the conflict situation.[5] The requirement to avoid or acknowledge conflicts of interest is in some respects comparable to the requirement to conduct an experiment in a double-blind manner (that is, so that the investigator as well as the human subjects are ignorant of which subjects are receiving the experimental treatment and which are the controls receiving a placebo). The requirement is not meant as a barrier to intentional wrongdoing, but because it is widely acknowledged that trying very hard not to be influenced by the knowledge of which patients are receiving the experimental treatment doesn't work. A sincere attempt not to be unduly influenced may lead to overcompensation. The lesser requirement of open acknowledgment is adopted for those conflict of interest situations in which it would be too burdensome or unworkable to require divestiture of other interests by the person in a position of trust. For example, some research journals require authors to disclose any substantial financial interests that might have biased their research assessment. Requiring investigators to divest themselves of investments that could be influenced by the results of their investigation would be too burdensome and might even discourage publication.

Dictionaries frequently apply the term "conflict of interest" only to conflicts between a person's *private* interests and that person's professional or official obligations and responsibilities. However, there can also be conflicts of interest in which private interests do not enter. For example, in the legal profession one is forbidden to represent both parties who are in a legal dispute, even if the lawyer is asked to represent one party only in a matter unrelated to the dispute between the two. This prohibition extends to representation of either party by other members of the same law firm. For example, if a law firm does tax law for company A and company B sues company A, no lawyer from the same law firm could represent company B in the suit. This stricture on lawyers is a result of the adversarial nature of the legal system and the role of lawyers as advocates in that system. There is no similar requirement on engineering firms to refrain from, say, building manufacturing facilities for two companies that directly compete in the market, although the engineering firm might need to be especially careful to keep confidential any proprietary information they learn in building the manufacturing facilities. This example illustrates the point that one needs to look carefully at the nature of a professional's or public official's obligations and responsibilities to know when conflicting interests become a conflict of interest; that is, when a situation that requires discretion to handle the actual or potential conflict fairly is a situation one is morally required to avoid altogether.

COMMISSION PAYMENT UNDER
MARKETING AGREEMENT [6www]

Inaka, a licensed professional engineer, has been extensively engaged in engineering activities in the international market. Inaka intends to draw on this experience and on personal contacts to represent U.S. firms who wish to practice international engineering but who lack either a sufficient background in the special fields of knowledge required for that purpose or the resources to develop the necessary skills needed.

Inaka drafts a marketing agreement to address the need of many U.S. firms to develop their overseas potential without a large expenditure. Inaka proposes to develop contacts within stated geographical areas, evaluate potential projects, coordinate project development, and arrange contract terms between the client and the represented firm. For these services, Inaka will be paid a basic fee plus a retainer fee. Both fees will be negotiated on an individual firm basis. Inaka will also receive a marketing fee, which is a negotiated percentage of the fees actually collected by the firm for projects that Inaka successfully markets.

Is it ethical for an engineering firm to enter into a marketing agreement that agrees to pay a percentage of fees collected for projects Inaka marketed? Is Inaka's engineering judgment in danger of being compromised by desire to obtain the commission?

In making a judgment on the case that was the basis for "Commission Payment Under Marketing Agreement," the NSPE's Board of Ethical Review judged it to be a violation of professional ethics for an engineer to be paid a contingency fee based on the projects the engineer success- fully markets. They had made a judgment in a somewhat similar case (62-4) that it was ethically acceptable for an engineering firm to pay a combined salary-commission to a marketing employee who was not an engineer. They saw it as crucial that the person doing the marketing on commission was not offering a professional engineering opinion that could be influenced by the incentive of a commission. The Board stated the concern that ". . . this method of compensation is undesirable since it could lead to loss of confidence by the public in the professional na- ture of engineering services." These examples illustrate the point that ethical concerns and the moral rules formulated to address them vary from the practice of one profession to another.

The practice of giving or receiving a commission or contingency pay- ment for one's services receives attention in many provisions of past and present versions of the NSPE code of ethics. One concern is to prevent kickbacks, that is, giving or receiving payment for referring work to an- other. For example, the present NSPE code of ethics says, under Canon 5 (against deceptive practice),

> b. Engineers shall not offer, give, solicit or receive, either directly or in- directly, any contribution to influence the award of a contract by public authority, or which may be reasonably construed by the public as having the effect of intent to influencing the awarding of a contract. They shall not offer any gift or other valuable consideration in order to secure work. They shall not pay a commission, percentage or brokerage fee in order to secure work, except to a bona fide employee or bona fide established commercial or marketing agencies retained by them.

Under Category 5 of Professional Obligations (on conflict of interest) the code reads:

> b. Engineers shall not accept commissions or allowances, directly or indi- rectly, from contractors or other parties dealing with clients or employers of the Engineer in connection with work for which the Engineer is respon- sible.

And under Category 6 of Professional Obligations (forbidding im- proper means of obtaining employment or advancement) it reads:

a. Engineers shall not request, propose, or accept a commission on a contingent basis under circumstances in which their judgment may be compromised.

Differences between the ethical codes or guidelines of two professions might be due to differences in the level of ethical concern between two professions, but they are likely to reflect differences in the conditions of practice that influence the ethical significance of particular actions. Good ethical codes or guidelines do not provide a random collection of ethical advice. As we saw earlier, codes provide guidance for addressing moral problems that arise in professional practice but are unfamiliar in ordinary life. The problems that arise in professional practice vary with the profession.

To understand the ethical significance of some action, such as how an offer of money or goods might corrupt professional practice, or how carelessness might lead to failure in some professional responsibility, one needs to understand both **the responsibilities of a profession and the conditions of its practice.**

Understanding the practice of a particular profession is necessary to understand why certain responsibilities are particularly important or certain types of acts are particularly injurious in that profession. The difference that the professional context makes is most vividly illustrated by rules in the ethical code of one profession that have *no* counterpart in others. We saw one of these in the stricture on lawyers against representing two legal adversaries. As another example, physicians are forbidden to terminate their relationship with a patient under their care without first referring that patient to another provider.

> This is in contrast to the view that being ethical just means acting in accord with a set of moral rules that specify what everyone should do or refrain from doing. The rules that are germane and the temptations one must guard against vary from one profession to another. Furthermore, the discharge of responsibility involves more than rule following.

Once having undertaken a case, the physician should not neglect the patient, nor withdraw from the case without giving notice to the patient, the relatives, or responsible friends sufficiently long in advance of withdrawal to permit another medical attendant to be secured.

– American Medical Association, *Principles of Medical Ethics*

There is no such generally recognized obligation of engineers to continue rendering service until a replacement engineer can be found.[7]
Here is an example of a specific moral rule that applies in medicine but
not in the engineering profession. Patients, unlike the clients of engineers, are much more vulnerable and likely to fear abandonment in their
vulnerable state. Part of what the public entrusts to physicians is this special vulnerability. Although one can imagine emergency circumstances
under which an engineer ought, other things equal, to ensure that the
necessary engineering services would continue to be available, tender
regard for a client's or employer's special vulnerability is not part of what
engineers are expected to provide. The provisions in ethical codes and
guidelines of professional societies are justified insofar as they provide
guidance on what is necessary and important to fulfill the trust placed
in members of a profession.

Not only are some moral rules more germane to certain professions
than to others, and some ethical provisions applicable only to certain
professions, but the criteria for the application of some category may
vary with the profession or with the customs of the larger culture in
which the profession is practiced. For example, the determination of
the sort of gift that exceeds appropriate hospitality varies with profession
and with culture. A twenty-five dollar cutoff was commonly considered
a limit for appropriate gifts and hospitality in U.S. engineering circles
in the 1980s. Twenty dollars is the current limit on gifts to American
military officers. (Some of these limits are discussed by the NSPE Case
76-6, Gifts to Foreign Officials. That case deals with an apparent request
that an engineer pay a bribe or extortion, rather than an offer of a gift
or bribe to an engineer.[8www])

As further examples, the United States recognizes a right of freedom
of speech and interprets this right more broadly than do other technologically developed democracies. In some countries it is considerably
easier to prove defamation. This is true in Australia, for example, and
the IEA code of ethics reflects that view in its stipulation that its members
shall "neither maliciously nor carelessly do anything to injure, directly
or indirectly, the reputation, prospects or business of others."

RESPONSIBILITIES, OBLIGATIONS, AND MORAL RULES IN STANDARDS OF ETHICAL BEHAVIOR

Rules of conduct, rights, and obligations specify what acts a professional
is ethically required or forbidden to do, such as: "Engineers should not

sign off on work that they have not checked" or "Surgeons should not operate on patients without obtaining their consent." These express some aspects of the ethics of a profession. However, checklists of rights and obligations or rules of conduct only set minimal standards for ethical practice. One need not make any complex judgments to recognize that signing off on unchecked work or operating without consent departs from good practice. If rights and obligations or rules about what acts to perform or refrain from performing were all there were to professional ethics, it would be a simple matter and hardly worthy of attention in college courses. Professional responsibility is such a demanding subject because the exercise of responsibility typically requires the exercise of discretion and consideration of many technical matters and matters of value – such as how safe is safe enough.

The statement of ethical obligations or rules of professional conduct provide some help in distinguishing malpractice from acceptable practice but not much help in differentiating good or responsible practice from minimally acceptable practice. Good practice is needed. In choosing an engineer or a health care provider, it is not enough to know that the engineer or provider will not be grossly derelict in his or her duty or engage in malpractice. What one wants is someone who will provide *good* professional services. The exercise of discretion required to achieve high standards in professional work and the synthetic integration of technical knowledge with concern for the foreseeable effects of one's work on human well-being demands problem-solving skills very much like those used to address design problems.

As philosopher John Ladd and sociologist Bernard Barber have argued,[9] responsible or trustworthy practice requires both that the professional be proficient and that she exercise proper concern (or "due care" as it is often called in legal discussions). Responsible practice is practice that is conducive to producing desirable results – for example, technology that operates safely and reliably and performs its intended functions well. Unlike rules of conduct, rights, or obligations, the specification of responsibilities states the ends to be achieved rather than the particular act to be performed or to be avoided.

In a complex society like ours it is not possible for anyone to master all areas of human learning, or even all the areas that directly bear on one's own well-being. In an age when knowledge of matters bearing on human welfare is complex, everyone must rely on those who have mastered other parts of that knowledge. Although some practitioners may not in fact be competent or may not show proper concern for the welfare of their clients or the public in general, *there are no good alternatives to having professionals behave responsibly.*

The advanced study peculiar to a given profession usually includes practical as well as theoretical study, such as an internship or work experience. Mastery of the specialized body of knowledge is the basis for professional judgment. Because those outside of the profession do not have the same practical and theoretical knowledge, they cannot adequately evaluate the professional judgment of someone in that profession (although those outside the profession may appropriately place constraints on that judgment or raise important questions about particular judgments). Therefore, members of the profession should have a major voice in establishing and revising or maintaining norms of professional practice, as well as for judging when the norms have been met. Engineering knowledge is required to fully evaluate the plans for the construction of a bridge, although some inadequacies, such as the bridge being too narrow, might be obvious to the layperson, just as a knowledge of surgery is needed to evaluate the quality of a particular operation.

The case of Peter Palchinskii illustrates how the extent to which professions have control over the projects they work on affects the ability of their members to exercise judgment and discretion and carry out professional responsibility.

THE CASE OF PETER PALCHINSKII

The case of Peter Palchinskii is told in Loren Graham's book, *The Ghost of the Executed Engineer: Technology and the Fall of the Soviet Union.*[10] Peter Palchinskii was a multifaceted and extremely talented engineer in the U.S.S.R. during the Stalinist era. Palchinskii frequently criticized government policy for such things as inattention to the health and safety of workers, as well as for short-sighted planning. Although he was a committed Marxist, he was charged with treason and executed for pressing what were simply the concerns of a responsible engineer for matters such as worker safety.

Subsequently, engineering education in the U.S.S.R. narrowed significantly, perhaps to lessen the chance that other engineers would recognize the broader implications of their work and raise criticisms that the government did not want to hear. Palchinskii's story and the subsequent changes in engineering education in the Soviet Union provide an example of how social and political context affect the character of professional education and the scope of professional competence, and hence the capacity of professionals to recognize problems and act in the public interest.

Another example of an engineer who died for his ethical concerns is Benjamin E. Linder.

THE CASE OF BENJAMIN E. LINDER

As an undergraduate studying Mechanical Engineering at the University of Washington, Benjamin Linder became intensely interested in the human consequences of engineering and the introduction of technology in undeveloped areas to meet human needs. After graduation in 1983 he went to Nicaragua to work as a volunteer under the sponsorship of the Nicaraguan Appropriate Technology Project. (The name "appropriate technology" is the term widely used for technology suited to the needs of small producers, rural and urban, especially in the developing world. [11]) In the spring of 1984, Linder joined a project to provide power to a rural area in the mountains of northern Nicaragua that had no reliable source of electric power. Without electricity, there were no electric lights necessary to hold evening classes or refrigeration for vaccines and other medical supplies.

A small-scale hydroelectric plant was feasible, but without electricity there were no machine shops or skilled mechanics. Plans were made to accomplish the construction by teaching local people how to build, operate, and maintain the plant themselves. Linder taught local people how to work with concrete and use hand tools, so by May of 1986 when the plant was operational, many peasants had new skills and several were fully competent to run and maintain the plant.

The plant was used to power a small machine shop and support a medical center with a refrigerator. Future plans included a saw mill, a carpentry shop, and facilities to make cement blocks, bricks, and roof tiles for the local area.

During the 1980s the *contras* were working to overthrow the Nicaraguan Sandinista government. Their strategy was to attack farmers, teachers, and medical workers in outlying areas to weaken the government. The *contras* had been especially active in the area where Linder was working. When an organization of American citizens living in Nicaragua brought suit in U.S. court to stop the U.S. government from funding the *contras*, Linder joined the suit. In his affidavit he said he believed that his life was endangered. The suit was unsuccessful, but Linder was committed to his work. Two years later he was killed by the *contras* while making rainfall and flow rate measurements.

In 1988, the IEEE SSIT Award for Outstanding Service in the Public Interest was awarded to Benjamin Linder for his "courageous and altruistic efforts to create human good by applying his technical abilities."[12]

The stories of Palchinskii and Linder together vividly illustrate how the larger society may fail to support the exercise of responsibility by engineers. Since, as we observed earlier, there are no good alternatives to having professionals behave responsibly, the general population has a strong interest in fostering legal and other supports for responsible behavior by professionals.

Although cases of engineers who were killed for their aspirations are rare, such cases dramatically illustrate the importance of general societal support for ethical behavior. The American Association for the Advancement of Science's Human Rights Program monitors human rights abuses against science professionals around the world. According to their records, engineers significantly outnumber physicians as victims of human rights abuses. Of course, some of these human rights violations are for political actions of the engineers rather than for their practice of engineering.

TRUSTWORTHY PROFESSIONAL PRACTICE

For behavior to be responsible or trustworthy is a matter of both ethics and competence. It requires sustained attention to the well-being of others and the knowledge and wisdom to promote or safeguard those aspects of the other's well-being. In particular, to do a responsible job, a structural engineer who is building a bridge or a tunnel needs to have a concern for public safety and an understanding of public convenience, knowledge of traffic demands and environmental implications of the work, as well as a proficiency in structural design and sound estimates of the likelihood of earthquake and hurricanes in the locale. In overseeing the storage of chemicals, a chemical engineer or chemist must consider how the chemicals will react with one another if they leak or spill, as well as how the chemical will itself interact with various types of containers.

Ethical and technical considerations frequently become inextricably intermingled in the exercise of professional discretion and judgment. An engineer might never think to ask some particular technical question about the behavior of some material under unusual temperature or humidity conditions if that engineer has not first recognized the need to ensure the safety of device operators who may be working under very different conditions from the engineer's own. It is only after noting the potential for one device to influence another, as microwave ovens can influence pacemakers, that the question of how best to warn users of the potential risk becomes an issue. When the complex intermingling of ethical and technical issues is ignored, professional responsibility is treated as having separable ethical and technical components, and the exercise of responsibility is distorted. Of course, one *can* distinguish between reservations about a person's technical competence and concerns about his moral character in considering whether to trust the person's professional judgment. When he comes to exercise that judgment, however, technical and ethical components will usually be inextricable. Recognition of and concern about ethical questions commonly lead to new technical questions, and technical insights or breakthroughs often raise new ethical questions.

For a period, the bioethics conmunity discussed diagnosis as though it were an ethically neutral technical inquiry. Ethically interesting issues were assumed to enter only with the selection of treatment. It is now recognized that the question of whether to obtain more technical information in pursuit of a firm diagnosis raises ethical issues. Suppose a patient is seriously ill with a disease that will kill her within a year, and she is having symptoms of what may be a passing condition or may be the early stages of another disease with a slow progression. Are the pain, risk, disruption of life, and expense to the patient of tests to diagnose the new symptoms justified by the difference the knowledge will make to his treatment? In other cases one should consider whether patients are prepared to cope with unexpected information that may turn up – for example, when information about abnormalities with largely unknown prognosis turns up in prenatal tests such as amniocentesis.

One of the NSPE BER cases in the introduction used to illustrate the professional responsibility for safety also illustrates the interconnection of ethical and technical questions.

TECHNICAL DISAGREEMENT AND AN
ETHICAL RESPONSIBILITY [13www]

Hillary is an engineer working for the state environmental protection division. Hillary's supervisor, Pat, tells Hillary to quickly draw up a building permit for a power plant and to avoid any delays. Hillary believes that the plans are inadequate to meet clean air regulations, but Pat thinks that these problems can be fixed. Hillary asks the state engineering registration board about the consequences of issuing a permit that goes against environmental regulations and finds that one's engineering license can be suspended for such action. Hillary refuses to issue the permit, but Hillary's department authorizes it anyway.

Was getting an opinion from the state engineering registration board a responsible action on Hillary's part? Was refusing to issue the permit?

What other information would you like to have and what difference would it make to your assessments?

It is sometimes held that the ethical questions are questions of whether to do something; the question of how to go about it is merely pragmatic. As was discussed in Chapter 1, however, questions about how to do things often raise ethical questions of fairness, and questions of how far to go, say in protecting safety, are at the core of professional responsibility.

WHICH MISTAKES ARE CULPABLE?

Everyone makes mistakes. Some are trivial, many are regrettable, but the gravity of the consequences does not determine the extent to which a mistake is morally blameworthy. The term "honest mistake" is used for a mistake to which no blame or guilt is attached. Consider the mistakes that led to the injury and death of patients treated with the Therac-25.

LETHAL TREATMENT: THE THERAC-25 X-RAY MACHINE

The Therac-25, a radiation therapy machine, killed or injured patients at several North American health care facilities between June 1985 and January 1987.

When the technician operating the Therac-25 made a typographical error in entering instructions and tried to correct this mistake by using the delete key, the filter on the machine dropped away. The result was that the patient undergoing radiation treatment received a massive dose of x rays. Several patients were injured or killed as a result before it was realized that the machine was dangerously defective.

The Therac-25 had been poorly designed and inadequately tested. The story is a complicated one that highlights many subtle as well as gross mistakes. In particular, the design and testing of the linking of the hardware and software were totally inadequate. Competitive machines had a shield that would engage if the power were at a high level. Furthermore, management decisions in the face of evidence of safety problems varied from short-sighted to negligent.

Atomic Energy of Canada, Ltd., the company that made the device, had many problems and has since gone bankrupt. [14]

Although there were instances in which the operating technicians used poor judgment, the mistakes made by the technicians who simply made typographical errors that caused the patients to receive massive doses of radiation were honest mistakes; typographical errors are the sorts of mistakes that humans make even when they are attentive. Furthermore, the way in which the technicians attempted to correct their error was entirely reasonable, and they would have had no way of knowing the disastrous consequences that would result. It was an honest error [15] on their part. What negligence there was came in the design and testing of the Therac-25.

Not surprisingly, many of the technicians who operated the Therac-25 machines and whose errors caused the deaths or injuries were emotionally devastated by the realization that their error had caused death or serious burn or injury to their patient. Philosopher Bernard Williams has discussed this often-observed psychological reaction of "agent regret." [16] It is important to distinguish between doing something that is morally blameworthy and the normal psychological reaction of blaming oneself when great pain and suffering result from one's innocent actions.

Notice that "negligence" is applied as a term of negative moral judgment. It is applied to mistakes (what would be negligent other than a mistake?) that are morally blameworthy. Carelessness shows inadequate attention to something. Only if one is morally obliged to do something is one negligent for failing to do it. For example, if I forget to water my house plant and it dies, that is careless of me (I did not give it the care it needed to survive) but my act is negligent only if I had a moral responsibility to give that care. Moral responsibility is not ordinarily a part of someone's relationship to his house plants.

Whereas a negligent act shows insufficient care in a matter in which one has a moral responsibility, a stupid mistake shows a lack of judgment, and incompetence shows a lack of knowledge or experience. A careless person is one whose care and attention cannot be trusted. A stupid person is one whose judgment cannot be trusted; an incompetent one lacks some relevant knowledge – a person may be competent in one area and incompetent in another. Because of their special education and training, professionals are expected to achieve a higher standard than the average citizen in matters related to their area of professional knowledge and practice. They have moral responsibilities for matters of great human consequence, such as public safety. Furthermore, they are morally obligated not to take work that lies beyond their competence. Therefore, an engineer or other professional can be considered negligent for actions that would seem simple ignorance if performed by nonengineers.

People, not only careless or stupid ones, may do careless or stupid things under circumstances that put unusual demands on their judgment or attention. These lapses may be more or less excusable. To say that an act is excusable is to say that there were special features of the situation: what are called "extenuating circumstances." Extenuating circumstances show that although the act was not a good thing to do (was not justified), the act was in part the product of special circumstances rather than a flaw in the agent's character. So if a person had been drugged or was distracted by a major personal tragedy, she might be excused for even an act of gross negligence or stupidity. If, however, she is partially responsible for creating the special circumstance, the excuse has less force. Thus, if people do something stupid or negligent because they are drunk, drunkenness does not excuse their behavior. Being overly tired or in a rush to meet a deadline is a frequent cause of carelessness in engineering practice and research. How far tiredness and being rushed may excuse a mistake depends on the extent to which the person who is tired or rushed brought these circumstances on himself.

So far we have established that although people recognize some mistakes as innocent or "honest" mistakes, others are blameworthy. The extent of blame and the possibility that the mistake is excusable (and so should not reflect on the agent's character) depend in part on the extenuating circumstances. They also depend on how egregious the mistake is, that is, whether it shows extraordinary stupidity or a disregard of others' welfare. The philosopher John Austin illustrated this point by contrasting the plea that one had trod on a snail by mistake with the plea that one had trod on a baby by mistake. One is morally required to use more care in dealings with babies.

There is a legal counterpart to morally blameworthy negligence in the notion of criminal negligence. Professionals can lose their licenses for negligence, or even be prosecuted. Recently a physician was tried for murder after attempting a procedure used for abortions in the second trimester of pregnancy on a woman whom he knew or should have known was in the third trimester, killing her. In rare cases, engineers have even been criminally prosecuted for negligence in complying with a law, as the following case illustrates.

> To say that an act is justified is to say that it was the right thing to do in the circumstances in question, even though it might be the sort of act that would ordinarily be wrong. For example, suppose that Leslie suddenly knocks Pat to the ground. Ordinarily that would be the wrong thing to do, but it would be justified if Leslie saw that a piece of machinery was swinging toward Pat and might severely injure Pat. In contrast, one would speak of the act being excused only if it were not justified (that is, a good thing to do in the circumstances). For example, Leslie might be excused for slipping and falling into Pat.

PROSECUTION OF THREE ENGINEERS FOR NEGLIGENT VIOLATION OF RCRA

In 1988, Carl Gepp, William Dee, and Robert Lentz, three chemical engineers at the U.S. Army's Aberdeen Proving Ground in Maryland, were criminally indicted for violating the Resource Conservation and Recovery Act (RCRA), which the U.S. Congress had passed in 1976. All three were civilians and specialists in chemical weapons work. At issue was the storage, treatment, and disposal of hazardous wastes at the chemical weapons plant, the Pilot Plant where all three worked. Although they were not the ones who were actually performing the illegal acts, they were the highest-level managers who knew of and allowed the improper handling of the chemicals. Each defendant was charged with four counts of illegally storing and disposing of waste.

They were tried and convicted in 1989. William Dee was found guilty on one count of violating the RCRA. Robert Lentz and Carl Gepp, who reported to Dee and Lentz, were found guilty on three counts each.

Among the violations observed were:

... flammable and cancer-causing substances left in the open; chemicals that become lethal if mixed were kept in the same room; drums of toxic substances were leaking. There were chemicals everywhere – misplaced, unlabeled, or poorly contained. When part of the roof collapsed, smashing several chemical drums stored below, no one cleaned up or moved the spilled substance and broken containers for weeks. [17]

In their defense the three engineers had said that they did not believe the plant's storage practices were illegal and that their job description did not include responsibility for specific environmental rules. They were just doing things the way they had always been done at the Pilot Plant. [17]

Another example of gross negligence is to be found in the story of A. H. Robbins's design and marketing of their intrauterine device ("IUD," a type of contraceptive device), the Dalkon Shield. An IUD is implanted in the uterus and thereby prevents conception or implantation. To facilitate removal, all IUDs have some sort of string attached to them.

THE WRONG STUFF -- THE DALKON SHIELD

A. H. Robbins, the makers of the Dalkon Shield, had first used multifilament polypropylene strings. This was a reasonable choice, since this material was used for some surgical stitching. However, the polypropylene was somewhat stiff and sometimes caused penile trauma to the women's partner during sexual intercourse. Robbins then substituted a string with a sheath made of nylon 6 and fibers made of nylon 66. This material was similar to fishing line and was a very poor choice of material for the human body since it decomposes in such an environment. Worse yet, the area within the sheath provides an optimal environment for the culture of anaerobic bacteria. These would multiply within the sheath and burst out, causing massive pelvic inflammation, sterility, and perhaps even death. Robbins eventually went bankrupt under the pressure of the liability judgments against the company. [18]

Another difference between innocent and blameworthy mistakes is that it matters whether a person has made the same mistake previously; people of normal intelligence are expected to learn from their mistakes. This is a point that Charles Bosk makes powerfully in his book, *Forgive and Remember*, with examples from the training of new surgeons. Sometimes, as in many of the cases Bosk discusses, a mistake once made is never repeated. At other times, however, what we learn is that we are prone to a certain kind of mistake, so we add safeguards to prevent making them.

THE AUTONOMY OF PROFESSIONS AND CODES OF ETHICS

MAINTAINING PROFESSIONAL STANDARDS: LETTERS OF RECOMMENDATION [19www]

Niemeyer is an engineer working for a medium-size manufacturing company and is being considered for a promotion. Niemeyer's employer contacts other engineers who had worked previously with Niemeyer for their comments. One of these was Singh, who was currently employed by another company and who did not have any current direct professional relationship with Niemeyer. Singh replied to the employer that he would not submit a comment on Niemeyer's qualifications or engineering competence because Niemeyer had dropped his membership in the state professional engineering society. Singh stated that in his view all engineers have an obligation to support their profession through membership in the professional organization. Niemeyer alleges that Singh acted unethically in submitting that reply to the employer.

What is the basis for the obligation to review or comment upon a colleague's work? How stringent is the obligation, that is, is it difficult or easy to find considerations that override that obligation?

Are engineers who fail to participate in their engineering societies undeserving of such effort from their colleagues?

In general, what are the extent of an engineer's responsibilities for maintaining the profession, professional organizations, or professional standards? What, if any, sanctions are appropriate to use against an engineer who fails to live up to this standard?

Although those outside a profession may possess some aspects of expert knowledge – for example, nurses frequently know enough about some medical procedures to recognize some subtle mistakes made by a physician [21] – it is members of the same profession who are in the best position to evaluate one another's performance. This is the rationale for so-called **professional autonomy:** the control (regulation and oversight) by professionals of the work of their members.

Silly as well as sound reasons have been given for believing that professions should be autonomous. For example, it is sometimes falsely alleged that professionals are inherently more moral than others. This view was more popular when morality and economic status were often assumed to vary together. The term "gentleman," for example, is ambiguous whether it means a man with a particular set of virtues, as in "a gentleman keeps his word," or only a man of a given economic and social status. The term "professional," like the term "gentleman," carries prestige and connotations of relatively high socioeconomic status. Not everyone who has such status lives up to a high moral standard.

Just as professionals vary in their demonstration of responsibility, so professions and professional organizations may do a better or worse job of overseeing the ethics and competence of members of that profession. Sociologists have written extensively about the self-protective behavior of professional organizations, and the self-serving character of some provisions within some codes of ethics, but codes of ethics may also represent a codification of the ethical norms of a profession and an effort to see that adherence to those norms is widespread among its members.

Codes of ethics in previous decades frequently contained provisions that contravened other ethical norms, such as prohibitions against criticizing the work of another member of the profession. Codes of professional ethics even today may contain provisions that lack ethical import. An example from a recent but outdated version of the NSPE Code of Ethics is a prohibition against engineers advertising their professional practice with "slogans, jingles, or sensational language or format." [22] This clause replaced one in the 1974 revision that had prohibited engineers from any advertising. [23] A major revision of the NSPE Code had occurred after two Supreme Court decisions in the late 1970s. In 1977 in *Bates v. State Bar of Arizona*, the Court ruled against the Arizona Bar association for attempting to prohibit advertising by two lawyers. The Court said that such prohibition violated the Sherman Anti-Trust Act. In April of 1978 the Supreme Court struck down Section 11c of the NSPE Code of Ethics that had prohibited engineers from engaging in competitive bidding, saying that this, too, was restraint of trade.

Earlier codes of many professional societies, including those of engineering and medical societies, had often forbade practitioners to criticize the work of other practitioners. Those provisions were a more serious impediment to the ethics of the profession than prohibition on advertising and competitive bidding, because they effectively undercut self-regulation by those professions. I know of no current code that still contains a prohibition on the criticism of peers, except for some that prohibit biased or unfair criticism. For example, in the current (1996) code of the NSPE we find:

> Engineers shall issue no statements, criticisms or arguments on technical matters which are inspired or paid for by interested parties, unless they have prefaced the comments by explicitly identifying the interested parties on whose behalf they are speaking, and by revealing the existence of any interest the engineers may have in the matters.
>
> – item 3c under Rules of Practice

> Engineers shall not attempt to obtain employment or advancement or professional engagements by untruthfully criticizing other engineers, or by other improper or questionable methods.
>
> – item 6 under Professional Obligations

> Engineers shall not attempt to injure, maliciously or falsely, directly or indirectly, the professional reputation, prospects, practice or employment of other engineers. Engineers who believe others are guilty of unethical or illegal practice shall present such information to the proper authority for action.
>
> – item 7 under Professional Obligations

The closest to a general prohibition on the criticism of other engineers is a prohibition on "indiscriminate criticism" found in the ASME code.

> Engineers shall not maliciously or falsely, directly or indirectly, injure the professional reputation, prospects, practice or employment of another engineer or indiscriminately criticize another's work.
>
> – item g. under canon 5, enjoining engineers not to unfairly compete with other engineers.

The members of professions do sometimes abuse their power by making the status and control of their profession their exclusive concern at the expense of concern for the good of the public or their clients. Such actions eventually lead to public distrust, but there is no simple way to prevent such abuses of power. Professional judgments are generally best monitored and evaluated by other professionals. This is the justification for the autonomy of professions, that is, the conclusion that professions must evaluate and have a central role in regulating their own practice. Professions may fail in control of their members, however, and such failure usually leads to greater regulation.

Given the history of professional codes of ethics, the question of whether the code of a particular professional society reflects the best moral reasoning of its practitioners is an empirical one; one does not know that a code of ethics actually has any ethical content before examining the code. As we saw in the introduction, some codes of ethics list as one of their principles statements such as "Engineers uphold and advance the integrity, honor, and dignity of the engineering profession by using their knowledge and skill for the enhancement of human welfare," which shows an overriding concern for the status of their profession. Enhancing human welfare is represented only as a means to that end.

John Ladd has argued against the uncritical acceptance of codes of ethics as authoritative ethical guides in "The Quest for a Code of Professional Ethics: An Intellectual and Moral Confusion."[24]

In the past two or three decades, the codes of many engineering societies have been reformed. For the most part, they articulate the ethical requirements on engineers to which most of the membership of those societies can subscribe. However, prohibitions such as that against advertising one's professional practice with jingles,[25] or against joining a strike,[26] illustrate that considerations of the image and status of the profession find their way into these codes.

In engineering ethical codes and guidelines, and especially in some of the decisions by the NSPE's Board of Ethical Review, the treatment of one's fellow engineers and concern for the dignity of the profession receive strong, sometimes surprisingly strong, emphasis. For example, consider the following case.

INFORMATION DUE POTENTIAL PARTNERS [27www]

Armandi, a principal in an engineering firm, submitted a statement of qualifications on behalf of his company to a governmental agency for a project. Armandi was notified that his firm was on the "short list" for consideration along with several other firms, but that it did not appear to have qualifications in some specialized aspects of the requirements, and that it might be advisable for the firm to consider a joint venture with another firm with such capabilities. Armandi then contacted Bent, a partner of a firm with the background required for the specialized requirements, and invited Bent's firm to join in a joint venture if Armandi was awarded the job. Bent agreed.

Thereafter, Engineer Chou, a principal in a firm that was also on the "short list," contacted Engineer Bent and also asked if the Bent firm would be willing to engage in a joint venture to supply the specialized services, if the Chou firm was selected for the assignment. Bent agreed but did not notify either Armandi or Chou of the agreement with the other.

The question that the NSPE Board of Ethical Review considers for all of its cases is whether the engineers, or at least the engineers who did anything unusual in the circumstances described, behaved ethically. (They base their judgment solely on the provisions of the NSPE code, sometimes bemoaning the lack of a provision that would forbid some action that they judge on independent ethical grounds to be wrong. Some changes in the two revisions to the NSPE Code adopted since the end of the 1970s have been made in light of the experience of the Board of Ethical Review.)

In its judgment on the above case the Board of Ethical Review judged Bent to have behaved unethically in agreeing to participate in a joint venture arrangement with more than one other engineering firm without making a full disclosure to all the firms. They cite two sections of the then current version of the NSPE Code of Ethics as potentially relevant:

Section 1 – The Engineer will be guided in all his professional relations by the highest standards of integrity, and will act in professional matters for each client or employer as a faithful agent or trustee.

Section 8 – The Engineer shall disclose all known or potential conflicts of interest to his employer or client by promptly informing them of any business connections, interests, or other circumstances which could influence his judgment or the quality of his services, or which might reasonably be construed by others as constituting a conflict of interest.

They hold that in this case there is no potential or actual division of loyalty or conflict of interest since Bent's loyalty would be centered only with the firm that won the contract. Therefore, they judge that the disclosure requirement of Section 8 does not strictly apply in this case, because at this point Bent "does not have a 'client,' as such." However, they maintain that the relationship with each is a relationship of trust and that the call for the highest standards of integrity in Section 1 requires Bent to inform each firm of the agreement with the other. They hold that the relationship of trust with each firm should not be diluted by establishing a similar and possibly competitive relationship without disclosure of this fact to all parties; therefore, Bent behaved unethically.

Sometimes a failure to show consideration is a failure of professional responsibility. This is true, for example, when one's clients can be expected to be frightened, confused, and in pain, and so less than effective advocates for their needs. However, there is a point at which consideration, although virtuous, goes beyond what is morally *required*. Notice that Section 1 quoted above does specify dealings with clients (for those in private practice) or employers, rather than partners in joint ventures.

Many people regard telling both firms of one's promise to form a joint venture with the other as *praiseworthy* or a *good thing to do* (say, because it shows praiseworthy candor and honesty). But those with whom I have discussed this case usually find it excessive to say that such behavior is ethically *required*. Among the reasons for agreeing with them is that there is no implied understanding that an engineering firm would not make contingency plans and agreements if the conditions for one joint venture are not met, and entering into such a joint venture does not harm the firm that did not get the contract.

How should we understand the NSPE's view that a high level of consideration of one's fellow engineers is ethically required? Neither the code nor the NSPE Board of Ethical Review recommends that engineers ignore other ethically significant considerations to be considerate of fellow engineers. Rather they demand that engineers go to some extra effort to show consideration. Surely some consideration is required for engineers to form a cohesive community able to maintain standards of professional behavior. To maintain standards of behavior, competition must be carried out within a framework of standards of decency and fair play. Striking an appropriate balance between collegiality among engineers and competition that keeps engineers alert to standards of achievement and customer satisfaction is a delicate matter. Professions have sometimes erred on one side or the other.

On the one hand, some professions have suffered the loss of public trust when their members have put too high a priority on their loyalty to one another and sacrificed other morally significant aims for it. Physician peer review organizations, charged with overseeing the work of physicians, were found to be quite lenient with some physicians, notably those with drug and alcohol problems. Physicians who sat on the peer review organization committees feared that they would be personally sued for removing the license of a negligent physician. This fear contributed to the pattern of excessive leniency. (Such suits were common and proved to be a great burden to the physicians who were sued, even if the suit was ultimately lost.) To encourage better peer review, Congress enacted the Health Care Quality Improvement Act in 1986. This act gives legal immunity to the actions of peer review organizations if their judgment is based exclusively on the competence of a physician, rather than on such matters as membership or lack of membership in a professional organization. Similar legislation to govern review of the professional conduct and competence of engineers and scientists would strengthen the hand of their professional organizations in maintaining high standards of ethical behavior.

> The fear of defending a lawsuit may inhibit university officials from reporting when a faculty member has been dismissed for cause, including research fraud. It also inhibits journal editors from printing retractions when not all the authors of an article agree that retraction is warranted. Professional societies such as the American Association for the Advancement of Science have urged similar legislation to protect university officials and journal editors who act in good faith in such cases. These considerations illustrate some of the factors that may inhibit peer control of professional conduct and therefore interfere with autonomy of professions.

On the other hand, the failure to maintain standards of fairness and decency in the competition among members of some professions has led to the failure in other duties to clients or to the public. This point has been missed by some writers who dismiss duties to fellow professionals as having no ethical content and being matters merely of etiquette. These writers seem to be so concerned with the duties that a professional owes directly to the public (including their clients) that duties to members of the same profession seem insignificant. However, fairness and decency to members of one's profession are also ethical matters. Furthermore, when those ethical standards are not met, the conditions for provision of good services are often undermined as well. Later, when we consider fair credit in research, we shall see the negative effect on the production of trustworthy research when standards of fairness and decency cease to govern the relations among researchers.

DOES EMPLOYEE STATUS PREVENT ACTING AS A PROFESSIONAL?

Discussions of professional ethics frequently start from an outdated model of professional practice in which the practitioner is assumed to be self-employed and in a one-to-one relationship with a client whose welfare is at stake. Indeed, it has often been argued that being an employee rather than in private practice makes a professional accountable to the employer and therefore less able to uphold professional values. Those who hold this view argue that the professional status of engineering is compromised by the fact that the majority of engineers are employees rather than in solo or group practice.[28] (This is true of the majority of mechanical, electrical, and chemical engineers, but not of the majority of civil engineers. It is increasingly true of physicians who used to be "in private practice" but are now likely to be employees of HMOs, clinics, and other health care organizations. Most lawyers are also employees, at least at the beginning of their careers. Most teachers are employees throughout their careers.)

The association of employee status with a compromise in professional status has some special relevance for engineers in the United States (but not, for example, in Canada[29]), because U.S. engineers employed in industry have an "industry exemption" from the requirement that they be licensed. As a result, the majority of engineers working for industry in the United States (unlike engineers in private practice, and *any* lawyer, physician, or nurse) *have no license to lose* for incompetent or unethical behavior. However, as we saw in the case of the three chemical engineers who handled hazardous waste improperly at the Aberdeen Proving Ground, they are liable to other penalties.

Employee status of a profession can also change over time. Most of what in the writings of Hippocrates was considered the work of a physician would now be seen as nursing. Since nursing differentiated from medicine, however, most nurses of the past century as well as this one have been employees. Therefore, it is anachronistic to continue to require self-employment as a mark of the 'true' professional.

Being both a professional and an employee creates some special problems, even if most professionals are now becoming employees. First, professionals who are employees must adapt to others more than if they were in private practice. They must figure out how to discharge their professional responsibilities without being so critical of minor departures from their personal standards that they impede important work. Second, they must figure out how to present their case on important issues so that others are more likely to appreciate their point. Third, if those within the organization continue to disregard an important matter, professionals must make a judgment about whether and to what extent they should either breach confidentiality or "make trouble for the organization" by taking matters outside the organization. Finally, they must decide where to take the matter and how best to raise the issue, to both get attention to the issue and be fair to others.

DO ENGINEERS HAVE A RIGHT TO PROTEST SHODDY WORK AND COST OVERRUNS?[30]

Kim is an engineer who works for a large defense company. Part of Kim's job requirements is to review the work of subcontractors the company employs. Kim discovers that certain subcontractors have turned in submissions with excessive costs, time delays, or deficient work and advises management to reject these jobs and require the subcontractors to correct these problems. After an extended period of disagreement with Kim over the issues, management placed a warning in Kim's personnel file and placed Kim on three months probation with a warning about the possibility of future termination. Kim believes that the company has an obligation to ensure that subcontractors produce specified work and try to save unnecessary costs to the government. Finally, Kim requests an opinion from the NSPE Board of Ethical Review on the matter.

Should Kim continue to protest management's actions?

What are the ethical issues in this situation?

What is your assessment of the actions of Kim's superiors?

Is there any further information needed that would make a difference in your assessment?

In this case the facts are presented as clear and unambiguous (and so display the subtleties and difficulties in making a judgment on even a clear case).

The ways that engineers, nurses, and other professionals who have for decades been called upon to exercise professional judgment and responsibility in complex organizational contexts have met their responsibilities and the difficulties that they have had in doing so provide lessons for professional ethics about meeting responsibilities in complex organizational contexts. Physicians, for example, are increasingly the employees of health maintenance organizations and other health care facilities. Their professional organizations, like the American Medical Association, are now taking on the task that engineering societies have faced for generations, of upholding the standards of responsible practice by employee professionals.

Although many engineers practice as employees, engineers have also been in private practice. The issue of whistle-blowing for employee engineers has many parallels with cases in which an engineer must weigh professional responsibilities against the obligation to maintain client confidentiality or otherwise protect the client's interests. We have already seen one of these in the introduction.

THE RESPONSIBILITY FOR SAFETY AND THE OBLIGATION TO PRESERVE CLIENT CONFIDENTIALITY [31www]

The owners of an apartment building are sued by their tenants to force them to repair defects that result in many annoyances for the tenants. The owner's attorney hires Lyle, a structural engineer, to inspect the building and testify for the owner. Lyle discovers serious structural problems in the building that are an immediate threat to the tenants' safety. These problems were not mentioned in the tenants' suit. Lyle reports this information to the attorney, who tells Lyle to keep this information confidential because it could affect the lawsuit. Lyle complies with the attorney's decision.

What, if anything, might Lyle have done other than keep this information confidential? Which, if any, of those actions would have better fulfilled Lyle's responsibilities as an engineer?

What other information may be needed to make this decision?

CODE VIOLATIONS WITH SAFETY IMPLICATIONS [32www]

Lee, an engineer, is hired to confirm the structural integrity of an apartment building that Lee's client, Scotty, is going to sell. Through his agreement with Scotty, Lee will keep the report confidential. Scotty makes it clear to Lee that the building is being sold in its present condition without any further repairs or renovations. Lee determines that the building is structurally sound, but Scotty confides in Lee that electrical and mechanical code violations are also present. While Lee is not an electrical or mechanical engineer, he realizes that the problems could result in injury and informs Scotty of this fact. In his report, Lee briefly mentions the conversation with Scotty about these deficiencies, but he does not report the violations to a third party.

What is your evaluation of Lee's actions? Of Scotty's? Is there any information not stated here that would make a difference to your judgment?

Professional education in engineering and science has only recently begun to give attention to questions of how to make fair and effective complaints. It has only developed in recent decades as a part of engineering ethics and in the past few years in research ethics. Until very recently, medical education has also neglected teaching this moral skill, although for decades physicians have sent their patients to hospitals and needed to be advocates for them in those facilities.

Given the reliance of others upon the judgment of professionals, whether those professionals are employees or self-employed, the more important questions than whether certain professionals are employees are:

- how best to prepare professionals to cope with the potentially competing demands placed upon them so that they behave responsibly;
- how to create supports within and without the employing institutions to support a high standard of ethical behavior; and
- how to create laws and policies that further rather than frustrate responsible practice.

In the effort to find criteria for responsible engineering practice, for example, there is no better place to begin than by making explicit the criteria that reflective practitioners use in evaluating engineering practice. These criteria are not beyond criticism, but they are grounded in a practical understanding of the realities and possibilities of engineering.

Society must rely on members of a profession to judge their peers' exercise of professional judgment. Therefore, as new consequences of some professional work are recognized, members of that profession must consider whether such consequences can and should be controlled. Criteria for responsible behavior regarding these potential consequences can then be proposed and discussed. Expectations need to be established about the consequences that a professional should take into account.

As I pointed out earlier, not <u>all</u> matters of professional conduct require significant discretion. The responsibility to ensure some future state of affairs differs significantly from the obligation to perform some particular act. To fulfill such an obligation, one need only be able to perform the act and be conscientious enough to follow through. To behave responsibly, however, an agent must decide what acts are required to attain the desired state of affairs. Regulation and monitoring may limit abuse of professional power, but they cannot by themselves produce responsible professional behavior.

That those whose welfare is at stake *need* trustworthy performance from the professionals they consult only implies that if the professionals in question are not trustworthy, those whose welfare is at stake will suffer. It does not imply that professionals should always be trusted.

SUMMARY

This chapter has discussed the basis and scope of professional responsibility, the norms of professional conduct, and how these go beyond mere specification of what acts to perform or refrain from performing. Responsible or trustworthy professional practice requires making complex judgments based on mastery of an expert body of knowledge to achieve or safeguard certain ends. Although everyone makes mistakes, we have seen that professionals are expected to achieve a higher standard than the general public in the area of their practice. As a result, certain mistakes are seen as negligent or reckless and morally blameworthy. A profession is characterized by a special body of knowledge. Those who have mastered that body of knowledge are in a special position to recognize when it is being used competently and with due care. This is the basis of the claim of professions to be "autonomous," that is, self-governing. To live up to this considerable claim, many professional organizations issue codes of ethics specifying the norms of behavior required for ethical practice. Members of a profession, however, do not always live up to the task of maintaining high standards of practice.

Because professionals who are employees are accountable to their employers, it is sometimes held that they have less control over how they practice. But as the majority of professionals now work as employees, the distinction between professions on the basis of what percentage of their members are in private practice is rapidly disappearing.

NOTES

1. Barber, Bernard. 1983. *The Logic and Limits of Trust.* New Brunswick, NJ: Rutgers University Press.
2. References to the NSPE code are to the latest, 1996, revision. Reference to other codes are to the versions current as of January 1997. Full text of the engineering society codes of ethics discussed here may be found in the WWW Ethics Center for Engineering and Science.
3. The provisions of the codes of ethics of Canadian provincial engineering societies have the force of legal regulation. The Quebec society, the OIQ , has four mounted police who carry out the police functions of their work.
4. As Heinz Luegenbiehl has observed, the justification for areas of professional ethics, such as engineering ethics, is that moral problems arise in professional practice that are unfamiliar in ordinary life. See Heinz C. Luegenbiehl, 1983, "Codes of Ethics and the Moral Education of Engineers." In *Business and Professional Ethics Journal,* 2(4):41–61. Reprinted in *Ethical Issues in Engineering,* edited by Deborah Johnson. Englewood Cliffs, New Jersey: Prentice-Hall, Inc., 1991, pp. 137–154.
5. In formulating this definition I had the benefit of a conversation with Michael Davis, although we do not entirely agree on the definition of conflict of interest.
6. Based on NSPE Case No. 78-7.[www] The NSPE cases are also available in hard copy in volumes V, VI, and VII of *Opinions of the Board of Ethical Review,* Alexandria, VA: National Society of Professional Engineers.
7. One engineering code of ethics does mention withdrawal from service to a client, but it does not require ensuring that the client have the services of another engineer. The 1983 Code de Deontologies des Ingenieurs du Quebec states:

> 3.03.04 An engineer may not cease to act for the account of a client unless he has just and reasonable grounds for so doing. The following shall, in particular, constitute just and reasonable grounds
> (a) the fact that the engineer is placed in a situation of conflict of interest or in a circumstance whereby his professional independence could be called in question;
> (b) inducement by the client to illegal, unfair or fraudulent acts;
> (c) the fact that the client ignores the engineer's advice.
> 3.03.05 Before ceasing to exercise his functions for the account of the client, the engineer must give advance notice of withdrawal within a reasonable time.

8. The case was originally published in *Opinions of the Board of Ethical Review,* Volume V, Alexandria, VA: National Society of Professional Engineers, 1981, p. 11.

9. Ladd (1978) and Barber (1983).

10. *The Ghost of the Executed Engineer: Technology and the Fall of the Soviet Union.* Cambridge, MA: Harvard University Press, 1993.

11. More information about this appropriate technology is obtainable at the Appropriate Technology International (ATI) web site, http://devcap.org:80/ati/ atiback.html.

12. This account is based on that by Stephen H. Unger, in his book *Controlling Technology: Ethics and the Responsible Engineer*, second edition, New York: Holt, Rinehart, and Winston, 1994, pp. 43–48. Unger also recounts additional stories of other engineers facing extreme situations.

13. Based on NSPE Board of Ethical Review Case 92-4. The NSPE cases are available in hard copy in the volumes V, VI, and VII of *Opinions of the Board of Ethical Review*, Alexandria, VA: National Society of Professional Engineers.

14. Nancy G. Leveson and Clark S. Turner, 1993. "An Investigation of the Therac-25 Accidents," *Computer* (published by IEEE), July 1993, pp. 18–41; Helen Nissenbaum, "Accountability in a Computerized Society," *Science and Engineering Ethics*, vol. 2 no. 1. An abstract of the study may be found in the WWW Ethics Center for Engineering and Science.

15. The term "error" is reserved for mistakes in relatively simple tasks where the standards of correctness are fairly clear, require no expert knowledge, and leave little room for "judgment calls" (i.e., discretion – for example, spelling errors and dialing errors). However, the distinction is often blurred, so you will hear people speak of "errors in judgment."

16. Bernard Williams, *Shame and Necessity.* Berkeley: University of California Press, 1993, pp. 70–1, 93.

17. Weisskoph, Steven, 1989. "The Aberdeen Mess." *The Washington Post Magazine*, January 15, 1989. p. 55, quoted in Harris, C. E., Jr. and Rabins, M. J. *Engineering Ethics*, "Aberdeen Three" Introducing Ethics Case Studies Into Required Undergraduate Engineering Courses, C. E. Harris, Department of Philosophy, and Michael Rabins, Department of Mechanical Engineering, Texas A&M University, NSF Grant Number DIR-9012252.

18. Harris, C. E., Jr., and Rabins, M. J. loc. cit.

19. I thank Robert M. Rose, Professor of Materials Science and Engineering at MIT, for information on this case. Professor Rose served as an expert witness in the case against A. H. Robbins for information on the Dalkon Shield.

20. This case is loosely based on NSPE BER Case No. 77-7.

21. Recognition of this fact has led nursing educators to include much more attention to the issue of raising ethical concerns in the professional nursing curriculum than is found in the curriculum of most other professions.

22. Professional Obligations 3a. For a copy of two versions of the Code of Ethics that both contained this clause see the *Opinions of the Board of Ethical Review of the NSPE*, Volume V (1981) and Volume VI (1989).

23. This was Item a under Section 3 of the NSPE Code as revised in January 1974. The prohibition on advertising was followed by clauses permitting such means of identification as professional cards, signs on offices or equipment, brochures stating qualifications, and brief telephone directory listings. A copy of the 1974 revision of the code is printed on the inside cover of the *Opinions of the Board of Ethical Review of the NSPE*, Volume IV (1976).

24. This essay appeared in *AAAS Professional Ethics Project: Professional Ethics Activities in the Science and Engineering Societies*, edited by Rosemary Chalk, Mark S. Frankel, and Sallie B. Chafer, Washington, DC: AAAS Press. It has been reprinted in *Ethical Issues in Engineering*, edited by Deborah Johnson, Englewood Cliffs, NJ: Prentice Hall, 1990, pp. 130–136.
25. This rule remained in the NSPE code of ethics until the 1993 revision.
26. The prohibition against joining strikes remains even in the latest (1996) edition of the NSPE Code of Ethics.
27. Based on NSPE Case 80-4.[www]
28. See, for example, Barber (1983).
29. Unlike both Canada and the United States, Australia does not now have a general licensing procedure for its engineers, although there is a National Professional Engineers Register (NPER) administered by the Institution of Engineers, Australia, for professional engineers who meet special qualifications.
30. Based on NSPE Board of Ethical Review Case 82-5.[www]
31. Based on NSPE Board of Ethical Review Case 90-5.[www]
32. Based on NSPE Board of Ethical Review Case 89-7.[www]

3

CENTRAL PROFESSIONAL
RESPONSIBILITIES OF ENGINEERS

The body of knowledge that characterizes a profession enables its practitioners to foresee possibilities, to devise ways to achieve desirable results, and to avoid undesirable side effects. Specialized knowledge enables engineers and scientists to design interventions, devices, processes, or constructions and to foresee how those products, processes, and constructions will act or interact. The designers of these products are uniquely qualified to foresee and modify many consequences of their use or misuse. Prominent among such consequences are safety hazards.

HOW CRITERIA FOR PROFESSIONAL CONDUCT CHANGE

Expectations of responsible professions vary not only with profession but also with the profession's experience of accidents and failure. Henry Petroski, in his elegant book, *To Engineer is Human*, uses examples from civil engineering to richly illustrate his thesis that engineering commonly advances by learning from failure. In almost every case the failures Petroski discusses threatened human health and safety.

Roland Schinzinger and Michael Martin, noting the extensive and often unpredictable character of the influence of technology on human life (not only health and safety), have argued that technological innovation amounts to social experimentation. In recent decades, informed consent has emerged as the primary criterion for ethically acceptable use of experimental treatments on human subjects. Schinzinger and Martin have suggested adapting a similar standard for the adoption of new technology.

Schinzinger and Martin propose using "proxy groups" composed of people similar to those who will be greatly affected by new technology. [1] Their mechanism for obtaining consent would not be the same as that used with human subjects in experimental studies. This difference is not surprising since Schinzinger and Martin develop their analogy not between engineering innovation and clinical experimentation but between engineering innovation and use of experimental medical *treatment*.[2] The use of an experimental treatment is governed by standards of competent care and informed consent for care rather than the more stringent norms applied to clinical experiments. For example, formal informed consent procedures are used for very few nonsurgical treatments. For a blood test or for an x ray, it is assumed that patients who cooperate with the testing procedures consent. More stringent measures are required to obtain the consent of experimental subjects, including prior review and approval of the study and the procedures for protecting human subjects by the facility's institutional review board (IRB).

The ambiguous character of many moral problems and, therefore, the importance of the moral agent's ability to devise courses of action that will be robust in the face of surprises, which we discussed in the first chapter, have special relevance to the professional responsibilities of those in science-based professions such as engineering and health care. These professionals, because of their special knowledge and experience, are likely to be the first to recognize health and safety threats.

Often in this book I compare the responsibilities of engineers and applied scientists with those of health care providers. This comparison provides useful information for the engineers and scientists who will work on biomedical technologies or in medical research. More generally it helps us examine professional responsibility from both the client and practitioner perspective. Health care is particularly instructive because everyone has had experience as a client of health care, and because the health care professions, like engineering and applied science, are science-based professions. The subject of the engineer's responsibility for health and safety, which we consider in the next section, highlights one of the striking differences between engineering practice and practice in the health care professions: Engineers and scientists generally have little or no direct relationship with many of those whose safety and welfare depend on their actions.

THE EMERGING CONSENSUS ON THE RESPONSIBILITY
FOR SAFETY AMONG ENGINEERS

The engineer's responsibility for safety is a useful place to begin discussion of specific professional responsibilities, because it is both familiar and generally agreed upon. It provides an undisputed example of responsibility that illustrates more general features of professional responsibility. Examination of the general features of professional responsibility for an uncontroversial case will provide an understanding that can be used to clarify responsibilities that are newer or more controversial.

Engineering students are often taught that safety is their responsibility. "First make sure the system doesn't do what you don't want it to do – that's the safety issue; then make sure it does do what you want it to do – that's the performance issue." This admonition is remarkably similar to the admonition to physicians: "First, do no harm."[3]

Emphasis on the engineer's responsibility for safety is also found in the codes of ethics or ethical guidelines of many engineering societies. These codes specify that it is the engineer's responsibility to protect public health and safety. Indeed, five of these societies – American Society of Civil Engineers (ASCE), American Society of Mechanical Engineers (ASME), American Institute for Chemical Engineering (AIChE), National Society of Professional Engineers (NSPE), and National Council for Engineering Examiners (NCEE) – state that the responsibility for public health and safety is the engineer's foremost responsibility. Seven societies enjoin the engineer to report or otherwise speak out on risks to health and safety.[4]

Although recognition of the engineer's responsibility for safety is long-standing, the range of factors that an engineer is expected to consider is ever-expanding. Lessons about the consequences of previous design decisions accumulate and technological innovation continues.

UNANTICIPATED FACTOR, AUTO SAFETY[5]

You are a new engineer working as part of a design team for a large automobile manufacturer. The company is doing a major redesign of one of its product lines.

Your team is responsible for designing part of the frame of the new car. As part of the company's drive to make cars lighter and more efficient, your team is directed to make some of the structural members out of carbon fiber composites. The cross member that holds the rails of the frame apart was ideally suited for composite replacement.

You tested several different composite materials and lay-ups, and finally chose one that you had reason to believe would work. Several prototypes of the car were built, which you checked carefully. The design was approved and is just about to go into production.

Just today you found a problem with your cross member. A few inches of the cross member from a car that was winter-tested showed extensive cracking. After looking at the design, you realize that the cracked portion rests on the exhaust system. You conclude that the hot pipe in cold weather created thermal stresses and caused cracking.

What can and should you do and how do you go about it?

KNOWLEDGE, FORESIGHT, AND THE RESPONSIBILITY FOR SAFETY

Experience with the consequences of engineering design decisions has broadened the range of consequences that responsible engineers are expected to foresee and the range of factors that they are expected to consider in controlling those consequences.

Not only the number of factors but the kinds of factors has increased. Engineers' responsibility to ensure the safety of a device or construction in its intended use is now only the beginning of what they must consider to ensure safety. For example, automobiles are not intended to have collisions, but inevitably many do. The concern to reduce injury and damage resulting from automobile accidents is therefore recognized as part of the responsibility of automotive designers.

The responsibility for making technology safe under extreme conditions (for example, severe storms), under foreseeable misuses, or as affected by foreseeable mistakes is added to the responsibility for considering unintended but frequently occurring circumstances.

As discussed above, experience with the consequences of engineering design decisions tends to broaden the responsibility for safety. The engineer's responsibility to ensure that a device or construction is safe in its intended use and under normal conditions is only the beginning. The responsibility for making technology safe under foreseeable misuses is also related to the engineer's responsibility for safety in extreme but foreseeable circumstances. Attempting to block every possible harmful misuse may be paternalistic, inasmuch as one may thereby block important beneficial action. As an extreme example, society could ban the sale of knives because people often accidentally cut themselves.

A good illustration of a foreseeable misuse is the misuse of a carpenter's hammer by a farmer who lost an eye as a result.

INJURY FROM MISUSE OF A TOOL

A farmer used the carpenter's hammer for a job in which a ball hammer was the appropriate tool. The forged head of the carpenter's hammer had become work-hardened with use, making it brittle and hence more likely to shatter when striking an object harder than itself. When the farmer used the hammer to drive a pin into a clevis[6] to connect a manure spreader to his tractor, a chip had broken off the hammer and injured his eye. The farmer brought legal action against the hammer's manufacturer for the injury. The manufacturer's lawyer argued that his client was not at fault because a carpenter's hammer is not intended to be used for the job that the farmer was doing. However, it was well known that carpenter's hammers were used for a variety of tasks, and work-hardening is a well-understood metallurgic phenomenon. Moreover, the manufacturer had received several chipped hammers that customers returned for replacement. The court found against the manufacturer holding that it should have foreseen the kind of use to which the farmer had put the hammer and acted to prevent such injuries.[7]

We saw in Chapter 2 that having a responsibility to ensure some future state or condition, such as safety, requires more than simply carrying out some required acts (fulfilling predetermined obligations). To behave responsibly, an agent must decide what acts are required to attain the desired state of affairs. The special body of knowledge that characterizes a profession gives the practitioners of that profession an enhanced ability to foresee what combination of actions will produce the desired end. Engineering knowledge, both theoretical and practical, enables engineers to design devices, processes, and constructions that perform as required and are safe in foreseeable modes of operation and under foreseeable conditions. Increased foresight raises the attainable level of safety and adds to the complexity of moral responsibility experienced by the professionals involved.

The concern with safety has led to the development of many structured techniques for safety review (where such techniques are applicable), such as the review of safety in a chemical plant. One such technique is the hazard and operability (HazOp) analysis used by industrial chemists and chemical engineers. Although structured, this method is not a simple recipe or algorithm. Like methods to aid conceptual design, this one begins with brainstorming by a team leader and process experts (such as design engineers, process engineers, toxicologists, or instrument engineers) to identify potential hazards. The team then carries out a unit-by-unit, stream-by-stream analysis of possible hazards in the plant process.

Other structured approaches to the identification and control of hazards are event-tree analysis and fault-tree analysis. Event-tree analysis begins with an initiating event (a mistake) and explores the states to which that event may lead. Fault-tree analysis begins with a malfunction or accident and reasons diagnostically to the circumstances that might have caused the malfunction or accident to estimate the likelihood of such accidents in the future. The types of possible consequences considered include fires and explosions, toxic chemical effects, ecosystem effects, and harmful economic effects. Because such techniques are now part of the technical subject matter in science and engineering, I will not describe them at length here. That they have been developed, however, further illustrates the intimate and constantly developing relationship between ethics and competence in professional engineering practice.

In "Is Idiot Proof Safe Enough?" Louis L. Bucciarelli critiques idiot-proofing as an ideal in engineering design. He argues that by treating the public as idiots rather than educating them, such zealous protection ultimately makes members of the public more vulnerable to injury, because idiot-proofing excludes the users from knowledge of the workings of the technology, and thus they are less prepared to deal with the unknown and unexpected.[8]

Other considerations about legal liability are important for understanding the relation or lack of relation between ethical accountability and legal liability. The function of liability judgments in the U.S. legal system is not primarily to assign moral blame. Society's interest in obtaining care for injured parties, which in other countries is handled by social safety net measures, leads the U.S. legal system to look for sources of support ("deep pockets") as often as for guilt. Furthermore, society's interest in having safer products is sometimes manifest in court decisions intended to drive the standard toward greater safety rather than punish an agent who violated the existing standard.

THE KANSAS CITY HYATT REGENCY WALKWAY COLLAPSE AND THE 1979 AMERICAN AIRLINES DC-10 CRASH

Two famous accidents in the last two decades – the 1981 collapse of the walkway at the Kansas City Hyatt Regency and the 1979 DC-10 crash – have expanded the scope of considerations still further. They dramatically illustrated that designs that may be safe when constructed or maintained as specified may nonetheless create hazards indirectly by creating temptations for others to take unsafe shortcuts. In these two cases the unsafe shortcuts were in maintenance, fabrication of connections, and construction or production.

THE 1979 CRASH OF AN AMERICAN AIRLINES DC-10 IN CHICAGO

The 1979 crash of a DC-10 occurred just after takeoff. The left engine had ripped off its mounting just before liftoff. The 1979 DC-10 crash was the worst disaster in U.S. aviation history. The cause of the crash was found to be a ten and a half inch crack in the rear bulkhead of the pylon that attaches the engine to the wing. The pylons of other DC-10s were found to have similar cracks.

FAA investigation found the cracks to have been caused by improper maintenance at American Airlines and Continental Airlines. Rather than separating the engine and the pylons during maintenance, as recommended by the manufacturer, McDonnell Douglas, the crews had been removing and reinstalling the engine and pylons as a unit using a forklift. The case is complicated by the fact that the FAA had approved the use of the forklift in this way, on the condition that leather straps were provided for cushioning. The heavy aircraft components were liable to be misaligned, resulting in cracking of the rear bulkhead's flange. The maintenance shortcut, however, saved approximately 200 person hours per engine, a significant savings. The designs of comparable aircraft by competing airline manufacturers, Boeing and Lockheed, did not present the same temptations for unsafe maintenance.

This accident helped to make aircraft designers aware of the ways in which their designs *may create temptations to take unsafe shortcuts at other stages of manufacture or maintenance.* Therefore, this case added a new element in the evaluation of a design's safety.[9]

HYATT REGENCY HOTEL WALKWAY COLLAPSE

On July 17, 1981, walkways in the Hyatt Regency Hotel in Kansas City, Missouri, collapsed leaving 114 dead and over 200 injured. Many of the dead and injured had been attending a tea-dance party in the atrium lobby at the time of the accident. Some had been standing and dancing on the walkways that were suspended above the lobby floor at the levels of the second, third, and fourth floors. Connections supporting the ceiling rods that held up the second- and fourth-floor walkways failed. The fourth-floor walkway collapsed onto the second-floor walkway directly below. The third-floor walkway, which was offset from the other two, remained intact. It was the worst structural failure in U.S. history in terms of injury and loss of life.

Jack D. Gillum & Associates, Ltd. had been subcontracted to perform all structural engineering services for the project. Jack D. Gillum was president of that firm and the professional engineer for the project. He was also one of the principals of Gillum-Colaco, the consulting structural engineering firm for the project from whom Jack D. Gillum & Associates subcontracted their work. Eldridge Construction Company was the general contractor on the project. Havens Steel Company was the fabricator for the connections, working under a subcontract to Eldridge Construction. During January and February 1979, over a year before the collapse, Havens Steel Company changed the design of the rod connections that hung two walkways, one above the other, from a floor above. The original plans had a single rod used at each point of connection, passing through the first walkway, fastened with a bolt underneath, and extending to suspend the mezzanine level. Havens proposed to use two separate rods to simplify the assembly task and to eliminate the need to thread the entire length of the rods. (See diagram below.)

This change doubled the load on the lower bolts, which now supported the weight of two walkways. The excessive load ultimately caused a lower bolt to pull through the beam so that one walkway collapsed upon the one below precipitating its collapse.[10] (As originally designed, the walkways were barely capable of holding up the expected load and would have failed to meet the requirements of the Kansas City Building Code.[11])

The fabricator, in sworn testimony at the accident, claimed that his company had telephoned the engineering firm Gillum-Colaco, Inc.

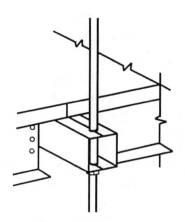

Figure 3.1. As Designed.

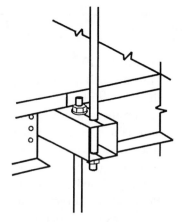

Figure 3.2. As Constructed.

for change approval. Gillum-Colaco, Inc. denied ever receiving such a call from Havens. [12] Yet, Jack D. Gillum had affixed his seal of approval to the revised engineering design drawings.

On October 14, 1979, while the hotel was still under construction, more than 2,700 square feet of the atrium roof had collapsed because one of the roof connections at the north end of the atrium failed. [13] The engineering firm testified that the owner, Crown Center Redevelopment Corporation, had on three separate occasions refused their request for on-site project representation to check all fabrication during construction because of the expense. [14]

Having learned from such accidents, engineers now more frequently consider how their design may indirectly increase risk. Following the Hyatt Regency disaster, the American Society of Civil Engineers (ASCE) urged civil engineers to accept design work only if they oversee the subsequent fabrication and construction as well.

Examples of other mistakes that threatened even graver consequences are those of the design and construction of the John Hancock Tower in Boston and the Citicorp Tower in New York City. Both structures had been designed in ways that met the standards of engineering practice of the day, but each turned out to be endangered by effects that were not considered at the time. The Hancock Tower case is summarized below. The Citicorp Tower case is discussed in detail in the next chapter.

UNEXAMINED INFLUENCES: BOSTON'S JOHN
HANCOCK BUILDING

The Hancock Tower manifests several unrelated problems. The excessive swaying of the tower, which LeMessurier sought to remedy with a tuned mass damper like the one he had pioneered for the Citicorp Tower, proved symptomatic of a serious structural problem. Structural engineering of the day did not take into account certain second-order effects that were significant for the John Hancock Tower. In particular, the increase due to gravity of the displacement caused by the wind hitting the building on its short side had been overlooked. This effect was greatly magnified because of the 300-foot length of the long side of the Hancock building. The correction of the problem for the Hancock Tower was a much more elaborate and expensive operation than the remedy of the Citicorp Tower's problems.

This case provides a stark lesson in the possible importance of effects and relationships that engineers may not know to examine.

The point that accidents are frequently the way in which a society learns to foresee dangers is thoroughly discussed in Henry Petroski's *To Engineer is Human*. Because of the human cost of learning through accidents and severe negative consequences, Schinzinger and Martin's recommendation that new technology be regarded on the model of experimental medical treatment is a telling one.

The history of engineering reveals at least two major criteria implicit in the evolution of what is required of engineers to fulfill their responsibility for safety. First, engineers' professional responsibility for safety extends only as far as the possible outcomes that an engineer can foresee at a given point in the development of engineering knowledge. Therefore, so-called end

> To be morally responsible for outcomes people must have some ability to foresee and influence them. I draw attention to this seemingly obvious point because some commentators have sought to blame technology (and engineers and scientists) for everything that is objectionable in modern life. Certainly technology has had harmful unintended consequences, as have political decisions. There is no possibility of doing without technology, however, any more than there is a possibility of doing without political institutions and decisions.

The view in philosophical ethics known as "Utilitarianism" maintains that what one ought to do in a situation is a matter of what, in those circumstances, would achieve "the greatest good for the greatest number." To know what one ought to do in a given situation would require a reliable estimate of the net amount of good produced by alternative actions. However, as we saw in Chapter 1, alternatives are rarely given, and instead must be devised. Indeed, for a problem that is technical as well as moral the responses that an agent can devise are strongly influenced by the agent's disciplinary background. So even if an agent were somehow able to know all the consequences of particular responses, the set of "possible responses" would vary from one agent to another.

Although Utilitarianism has these deficiencies as a guide to action, it may still be understood as one of several theories about what makes an action the best action, namely that it provide the greatest number, even if it is impossible to know whether a given action meets the criterion.

uses of technology (that is, the uses to which some technology is ultimately put), and social byproducts of technology, are the engineer's or scientist's *professional* responsibility only if the competent engineer or scientist can foresee them. Second, engineers have a professional responsibility to examine those determinants of results that they might control or influence.

In professional life, the responses one considers and the foreseeable consequences of those responses are a function of the state of knowledge at the time. Furthermore, the possibilities that a member of one profession will consider are necessarily very different from those that a member of another profession will consider when faced with the same circumstances. For example, faced with the threat of an epidemic of cholera, a civil engineer will think of sanitation improvements as a way to control the spread of disease. A physician will think of medical means.

In a democracy all citizens have some responsibility for the policies undertaken or allowed by their government, including policies on the development and use of technology. Engineers and scientists share this responsibility with other citizens even though they are not professional responsibilities peculiar to engineers and scientists.

HAZARDS AND RISKS

Thus far we have considered mainly those safety hazards that threaten to cause accidents and malfunctions. This was true of the hazards that structured techniques like hazard and operability study and fault-tree analysis are designed to identify, as well as the major accidents that are the subject of famous cases. The emphasis has been on identifying such hazards and bringing them to the attention of those with authority to eliminate the hazard.

Some safety hazards cannot be feasibly eliminated, or they can be eliminated or mitigated only by producing other adverse consequences. The comparison of the risks, in the technical sense of the degree or magnitude of harm, is facilitated by use of the techniques of cost-benefit or risk-benefit analysis. For example, the ingredient most effective in repelling insects, commonly known as DEET, is known to be toxic to humans. (The substance can also cause skin rashes and other adverse effects if it comes in contact with eyes or is ingested.) Nonetheless, people continue to use repellents containing DEET because less dangerous substances are less effective in repelling insects, and besides discomfort, insect bites may carry serious diseases like dengue fever and encephalitis. Use of DEET in some circumstances is justified by health considerations. The question becomes one of appropriate trade-offs between risks and benefits. It might make use of the risk-benefit analysis mentioned in the introduction on concepts, one of the structured techniques of risk analysis.

Risks that are tolerated in one area may not be tolerated in another, however. We have just seen a risk trade-off that is accepted for insect repellent. Trade-offs that are accepted for lotions and cosmetics are not tolerated for food additives. The standard of safety required for food additives is that their addition cause no harm to humans nor even cause cancer in other animals.[15] Furthermore, risks of additives are less tolerated than risks associated with substances that occur naturally. Many naturally occurring contaminants that cannot be entirely eliminated from various food stocks can produce significant harm. The presence of these contaminants has not led to banning the potentially affected food crops from the market.

We saw in the introduction on concepts that the notion of risk used in risk analysis and in risk-benefit calculations differs somewhat from the ordinary notion. Ordinarily when we consider the risk of a traffic accident, or of running out of water, or of tripping and falling, we focus on the negative event or situation that we would seek to prevent or eliminate. This notion is very similar to that of a hazard, except that a hazard is an externally caused threat: We may take a risk, for example by carrying only enough drinking water for the number of days we expect to hike, but we do not "take a hazard." We at most tolerate or accept a hazard.

Some notorious uses of cost-benefit analysis that trade safety for money have been generally condemned as irresponsible. Perhaps the best known example is Ford Motor Company's decision in the late 1970s about the Ford Pinto. After management realized the design of the Pinto left the car unusually vulnerable to explosion of the gas tank if hit from the rear, they decided against adding an inexpensive safety feature that would have lessened the risk. The decision was justified by a dubious cost-benefit calculation that assigned $200,000 as the monetary compensation for the pain and suffering of a burn death.[16] The Pinto gas tank case was loosely paralleled by the safety problems of the car design in the film *Class Action*, where those offering such techniques were derisively called bean counters. However, there were many features of Ford's use of cost-benefit analysis in this case that were both morally objectionable and prudentially short-sighted. Most would agree that death and injury due to a gas tank explosion is a hazard that must be made extremely rare before further reduction is traded off against other factors.

Trading off safety against cost considerations is not necessarily morally objectionable. It is more acceptable to limit protection against dangers that are to some extent under the control of those at risk. Consider the design of a new highway. The terrain may necessitate curves in the road. How much banking should those curves have? The greater the banking (the smaller the radius of curvature of the highway surface), the less likely will it be for cars to spin off the road, but the more expensive it will be to build. It seems reasonable to draw the line short of constructing the road as a speedway, but how big a safety factor should one build in? Is it enough to design for fair weather speeds of up to thirty miles over the speed limit? Forty? What about foul weather? What are the weather extremes in a normal year? In a typical hundred-year period? In a five-hundred-year period? How safe is safe enough, or, to put the matter another way, what is an acceptable level of risk? A judgment about the acceptable level of risk is implicit in any trade-off of safety against other considerations, or even of one sort of safety risk against another.

One way of lowering risks is to increase the safety margins or to "over design" in one's work. However, this is possible to different extents in various sorts of engineering work. Civil engineers normally build in a safety factor of 2, 3, or more. That is, they often construct buildings to withstand stresses two, three, or more times what they expect them to experience. If aeronautical engineers tried to use such safety margins, their planes would be too heavy to fly. Using much smaller safety margins, however, they must be all the more confident that they have not overlooked a potential cause of failure.

Industrialized societies have accepted the risk of airplanes designed with relatively small safety factors, because the actual rate of airplane accidents is acceptable to the flying public. In the United States the mortality risk in automobile travel is mile for mile greater than the risk of death in travel on a regularly scheduled airline trip.

In a society marked by rapid innovation, adverse consequences can at most only be conjectured. This is especially true regarding new chemicals.

SCENARIO: HOW DANGEROUS IS THIS CHEMICAL?

This weekend is the first time in a month that Pat has had some time to relax. Pat's team at Colortex had just come up with a new dye, a sulfated alpha-napthol, that promises to give Colortex a larger share of the commercial dye market. While doing some gardening, Pat recalls hearing that alpha- and beta-napthols are associated with high rates of bladder cancer and resolves to look into the matter on Monday.

On Monday Pat checks the data on the carcinogenicity of the napthols. At least the alphas don't seem to be as potent carcinogens as the betas, and a sulfate radical might make the chemical even more benign. However, Pat resolves to take up the matter with the team leader, E. D. Able. E. D. is rather cold and brusque, not an easy person to talk to. Pat does not look forward to raising a "nonstandard" issue with E. D. but remains concerned about the dangers of a potential carcinogen.

When Pat finally gets a minute with E. D., E. D. dismisses the issue, saying, "If we were going to worry about every ring structure, we'd have to test cholesterol for carcinogenicity. Life can't be made risk-free. Besides, assessment of carcinogenicity is the EPA's job. They will get notification 90 days before we market this dye. They can raise any objection then."

Of course, the EPA will eventually do its own assessment, but first many Colortex employees would be exposed to the chemical as Colortex moves from development to production of the dye. E. D. does have more experience than Pat, but Pat is not comfortable with the "innocent until proven guilty" stance that E. D. is taking toward this chemical.

What can and should Pat do?

THE SCOPE AND LIMITS OF ENGINEERING FORESIGHT

Engineers and scientists *are* likely to be the first to recognize many sorts of potential dangers to health and safety. However, they may not be particularly well qualified to either predict or control certain of the consequences of their work. For example, many societal effects, such as the growth of the suburbs, a reduction in the risk of tetanus (from the wastes of carriage horses), and some of the rise in lead levels in the environment, are attributed – at least in part – to the automobile. It is unreasonable to fault those who designed and developed automobiles for not foreseeing and mitigating all the negative consequences of the automobile, since we make no similar demands of other professions. The experience with the automobile teaches many lessons about control of technology, only some of which pertain to engineering. In a democracy all citizens in general, and not only engineers and other technical experts, are expected to consider and have a voice in major decisions about such areas of technology as transportation policy. As was pointed out earlier, this gives all citizens of a democracy some responsibility for the uses of technology within that society.

Which sorts of lessons from prior technology are the special responsibility of engineers to learn? If a negative effect *is* something that existing knowledge allows engineers to foresee, does that fact alone make the effect something that is the sole responsibility of engineers? An example will help to focus consideration of this question.

In an editorial, "A Sense of Sin," which appeared in the February 1988 issue of the *Biomedical Engineering Society Bulletin*, Steven M. Lewis argued that biomedical engineers are responsible for all the consequences of the devices they design and develop. Although the engineers' responsibility is clear enough in certain cases of safety problems under reasonably expectable use, operating conditions, and maintenance, Lewis claimed more. He contended that the engineers bore a special responsibility for the fact that life-support devices (respirators or ventilators, in particular) are often used inappropriately, such that their use merely prolongs patient suffering and wastes resources.

Now that society has experienced the harm resulting from inappropriate use of life-support technology, it is quite foreseeable that new life-support technology could be misused in the same way. Does this mean that it is irresponsible for engineers to develop any more life-support technologies? Notice that the knowledge on which the prediction of negative consequences would be based is not specific to engineering, but rather to North American society's reluctance to squarely face death and dying.

Lewis apparently believes that because engineers can now see that their work on medical technology may make it possible for this harm to be done, they therefore bear some responsibility for this misuse. (When I conducted interviews as part of a study of the responsibilities of biomedical engineers, I found biomedical engineers to be a rather diverse (one might even say maverick) group. Although many come to biomedical engineering through a series of fortuitous occurrences, many came to biomedical engineering because of a concern about the end use of their work. In some cases, this took the form of wanting to provide devices that would directly benefit people. In others, it took the form of wanting to avoid military work without leaving their technical area. I even found one respected biomedical engineer who took early retirement when he discovered that one of his principle inventions proved not to have been of any particular benefit to patients.)

Should devices such as respirators be taken off the market? Some types of devices have been removed from sale when hazards from certain uses are great. At one extreme lie semiautomatic assault weapons and semiautomatic rifles, pistols, and shotguns. The argument that their intended use – namely, to kill and maim people – is ethically unacceptable led to a ban on the manufacture of many classes of these weapons in the 1994 Crime Bill. [17www] A somewhat different example involves three-wheeled "dune buggies." These were intended to be a recreation vehicle but were removed from the market when it was found that children and adolescents driving these vehicles often had serious accidents.

The use of human growth hormone, somatotropin, as a drug treatment has been criticized because the inappropriate use poses unknown hazards to children. (The criticism has led to a call for greater controls, rather than complete removal from the market.) The hormone was commercially developed as an effective treatment for a form of human dwarfism caused by deficiency of the growth hormone. However, the use of the genetically engineered human growth hormone in the United States far exceeds the amount to treat the 2,500 cases of growth hormone deficiency found in children each year. It appears that some parents are subjecting their children (especially their male children) to unknown risks of side effects simply to increase the height of children who would otherwise be of normally short stature.

In the human growth hormone case, unlike the dune buggy case, the intended use of effective treatment for children who would be significantly shorter than normal is regarded as too important to forego. Therefore, there has not been a significant drive to ban the hormone altogether. However, serious arguments have been offered against developing treatments for traits that, although outside the statistical norm, are not painful or seriously debilitating to their bearer. One argument is that making these conditions the object of treatment contributes to stigmatizing them and the people who have them.

Life-support technology, such as the respirator, provides incontrovertible benefits to a great number of people, which is why no one has seriously suggested that we ban life-support technology. Furthermore, respirators and other life-support technologies are different from products that are difficult or arduous to maintain properly, but dangerous if not maintained correctly. As we saw in the last section, one example of this sort was the design of the DC-10 that contributed to the 1979 DC-10 crash. Another is a type of surgical drill used in cranial operations. It was designed to stop automatically after drilling through the skull bone. If this drill was not cleaned, maintained, and assembled properly, instead of stopping after drilling through the skull, it would sometimes plunge into the brain. The automatic stopping may have benefited some patients (for example, those with an unusual skull thickness), but the tendency to plunge did great harm to others. Resultant injuries have been the subject of lawsuits against everyone from the drill's manufacturer to the hospitals and the surgeons performing the operations.

Alternative design of aircraft or skull drills can reduce the likelihood of improper maintenance or the risks that result from it. In other words, by modifying their designs, engineers could remove or lessen the danger. Recall the earlier discussion of how engineers learn to consider new factors through learning from previous failure. When clothes' dryers first became popular consumer items, they would resume their operation when the door was reshut after opening in mid-cycle. As a result, some toddlers died or were severely injured after crawling into dryers that had stopped mid-cycle. Now safeguards to protect toddlers and young children are an essential consideration in the design of household appliances. In this case the safeguards are readily added to the design of the device, creating "an engineering solution."

In the case of hazardous waste, engineers now consider how to avoid using or creating hazardous substances in processes of manufacture and maintenance as well as how to control the release of hazardous substances and to provide means for safe disposal of such wastes. The creation of such processes is still a technical engineering task, albeit a relatively new one. The control of the clinical use of respirators is not.

Overusing respirators and other life-support technologies differs fundamentally from using a dryer as a playhouse or a dune buggy as a toy for children. What differentiates the appropriate use from a misuse of respirators is not always obvious; in contrast, playing house in a dryer is very different from using it for drying clothes. In the dryer case, engineers could find an engineering solution to the hazard posed to toddlers by making operation of the dryer conditional on throwing a switch that adults can reach but toddlers cannot. What differentiates relevant use and misuse of the respirator, however, is the prognosis of the patient with whom it is used. Because the physical actions taken with the respirators do not differ between use and misuse, misuse cannot be prevented by engineering means.

Lewis takes the phrase "a sense of sin" from J. Robert Oppenheimer's famous description of science as having "known sin" in producing the atomic bomb. Lewis's use of the phrase suggests that producing devices like respirators is akin to producing a weapon of mass destruction. It draws attention to the value implications of work in science and engineering. However, the design and development of life-support technology differ from the creation of weapons of mass destruction in morally significant respects. Weapons of mass destruction are devices that, if they function *as intended*, have overwhelming negative effects for some people and arguably for humankind. Used appropriately, life-support technology need not have negative effects for anyone.

Lewis is right about the suffering caused by the overuse of life-support technology, but in this case the responsibility for preventing or remedying the situation lies with parties other than biomedical engineers. As Stephen Lammers has pointed out, our society often looks to technology to eliminate the necessity of coming to grips with perennial human problems like death and vulnerability to disease and injury.

The example of the respirator illustrates the general point that although the possibility of foresight is required for engineers to be responsible for some harmful effect of technology, it also makes a moral difference if engineers are able to do something about the problem. The total configuration of consequences also makes a difference: If an engineer has good reason to believe that society is not able to appropriately control the use of some new technology or device and its effects will be wholly or predominantly harmful, that constitutes a good reason to refuse to work on it. If, however, the technology or device produces mixed results and only some uses are harmful, engineers have only the same responsibilities as other people for improving the social control of technology.

What examples do we have of such social control? In the few years since Lewis wrote his editorial there has been some progress on the evident problem of the overuse of life-support technology.[18] This progress exemplifies the diverse means societies have for controlling the uses of technology.

THE PATIENT SELF-DETERMINATION ACT

This national legislation took effect in December 1991. It requires hospitals and other health care facilities (nursing homes, hospices, home health agencies, HMOs, and other facilities that receive Medicare and Medicaid payments) to maintain written procedures on advance directives. Advance directives for health care are usually in the form of a living will or a health care proxy statement. A living will specifies in general terms the care that a patient would or would not want to receive if no longer able to consent to care. A health care proxy directive (also known as a "durable power of attorney" statement) specifies whom the person would wish to decide his medical care were he unable to do so.

When patients are admitted to a facility, that facility is required to notify patients that they have a right to refuse treatment and ask if they have made out advance care directives. Health care facilities are also charged with complying with state law regarding such directives.[19]

Most states have recently given legal recognition to living wills, health care proxy directives, or both. These state laws, together with the Patient Self-Determination Act, effectively address the problem Lewis raises, although none of them require engineering knowledge. Where, as in this case, engineering knowledge does not help either to foresee or to remedy some misuse of technology, engineers have only the same responsibilities as other citizens to prevent the misuse.

These considerations about the scope of an engineer's responsibility for safety suggest the following questions:

- Which of the possible societal consequences of technology can engineers (practicing in a country such as the United States in the last decade of the twentieth century) foresee?

- Under what conditions or to what extent can such engineers judge which influences and outcomes are desirable, and when is that determination better made by others?
- Under what circumstances would such engineers bear some responsibility to ensure those outcomes through their work (design, development, manufacture, etc.)?

When those questions have been answered, we still need to clarify how one goes about fulfilling such a responsibility when one does have it. As we saw in the first chapter, fulfilling responsibilities is a complex task, like a design problem. Even when the safety issue itself is relatively straightforward, the circumstances may make it relatively difficult for an engineer, especially one who is new to the job, to devise a good response. Consider the following scenarios based on experiences of recent engineering graduates:

IMPAIRED COWORKERS [20]

You are a student intern working second shift in the testing of a million-dollar spacecraft. Tasks on your shift are simple (regulating pressures, turning on or off different valves, reading off numbers from flow meters, etc.). However, keeping attentive and alert is very important, because turning a knob in the wrong direction could cause major damage to the spacecraft.

During your work break you go for dinner with two of the full-time employees. They pick out a bar/restaurant and order large (big goblet-sized) glasses of beer. You refuse when they offer you one. During dinner they each have another and wind up using more than the allotted hour break.

You are certain that they must have been affected by the amount of alcohol they drank, although you do not see anything amiss in the way they are operating the equipment. Drinking on the job is not allowed by the company, especially by those working near the spacecraft.

What, if anything can and should you do about the situation? Where might you go for advice?

NEW EMPLOYEE AND WORKPLACE SAFETY [21]

I have just received my Bachelors of Science in chemistry and started a job at Acme Wax Refinery. After three weeks on the job I have just completed my initial training from the head chemist at the plant. My job consists mostly of testing the purity levels of the various types of waxes. Because many of the company's customers are cosmetic manufacturers, we have to make sure the contents of our waxes conform with their industry standards of purity.

My first few weeks at this job also included a tour of the facilities, where I got a chance to meet the factory floor employees. These blue collar workers are the people that I will be working with day in and day out. There is some tension between the floor workers and the head chemist because the workers feel that they are constantly being stopped in whatever they are doing just to help the chemist get a sample of wax from the warehouse.

I have tried to strike up friendly conversations with them but to no avail. They have been quite unresponsive to my greetings and requests.

On the first week I began my work at Acme Wax, a call came in from a cosmetics company saying that the mineral content in one batch of our carnuba wax was too high. I was sent to the warehouse to get samples from the ones in inventory. When I got to the warehouse, I found that the crates of cosmetic grade carnuba wax were stored on palettes on the highest shelf, almost 20 feet high.

Realizing the rush on this job, I ask a floor worker on a forklift to lower the two crates so that I can take some samples. The worker thinks it is ridiculous to pull two heavy crates down just so I can take a couple of samples. The worker suggests, instead, that I get onto the forklift and he lift me to the top to gather the samples. Realizing the safety issues involved, I tell him that it would be a bad idea and ask him again to lower the crates down. At this point, the worker tells me not to waste his time.

What am I to do?

MATCHING AN ENGINEER'S FORESIGHT WITH
OPPORTUNITIES FOR INFLUENCE

The safety of the Chernobyl nuclear power plant had been criticized in Soviet technical journals. Before the 1974 DC-10 crash, engineers had warned that the design of the airplane was faulty. Preparing engineers to recognize safety hazards, although vitally important, is clearly not enough to prevent many accidents. Because engineers often recognize a hazard but do not have the authority to remedy it, and may be unable to get attention to the problem by those in their organization, engineering ethics has widely discussed whistle-blowing by engineers, that is, an engineer taking a concern outside her organization. However, whistle-blowing in this sense always marks organizational failure. [22] A better way of matching an engineer's influence with her insights and foresight (one that is less costly for all concerned), is for organizations to become more responsive to their engineers. Although this alternative has been the subject of much attention in industry, it has not been discussed to the same extent in other contexts.

There is a growing consensus among engineering organizations on the subject of raising safety concerns, both through complaining (within an organization) and through whistle-blowing (outside the organization). First, engineers have a right to force attention to many types of error and misconduct – such as waste and misrepresentation in work done under government contract – even by going outside the organization. Second, engineers have not only a right, but a moral obligation, to bring the matter to light when human life or health is at risk.

For some negative consequences, such as threats to public safety, an engineer is morally *obliged* to go to special lengths – including contacting outside agencies – to draw attention to the problem if that engineer is unable to influence the situation by the means offered within the organization. Of course, an engineer must exercise judgment about appropriate places to take a concern. (NSPE Case 88-6 recounts a case in which the NSPE Board of Ethical Review judges the whistle-blower to have handled the situation badly. [23www]) A government agency charged with regulatory oversight is usually an appropriate place to go. Engineers' experience in going to the press varies greatly. At a minimum it is difficult to describe technical matters in terms that a newspaper audience can understand. Furthermore, journalists are at least as subject to the temptation to sensationalize to sell newspapers as managers are to pay insufficient attention to safety to meet a deadline. Therefore, it is a good idea to know journalists and their motives before giving them sensitive information. (For a perspective from within the engineering profession on the appropriate use of the media in raising ethical concerns, see the NSPE Board of Engineering Review discussion of Case 88-7. [24www])

SUMMARY

This chapter has discussed the central professional responsibilities of engineers and scientists. The standards of responsible professional practice are not static but change as new knowledge is acquired. However, at least among engineering groups, there is a high degree of consensus that the responsibility for safety is foremost among the professional responsibilities of engineers. The scope of factors that can affect the safety of technology is unlimited. Some of these factors are discovered only through accidents, but the domain of facts that an engineer is expected to consider continually expands. The standards of responsible practice at any given time depend on what engineering knowledge of the time permits one to foresee and influence. It is in everyone's interest that engineers be heeded when they foresee risks and threats to the public welfare.

NOTES

1. Martin, Mike W. and Roland Schinzinger, *Ethics in Engineering*, second edition, New York: McGraw-Hill Publishing Company, especially pp. 63–78.
2. Ibid., p. 68.
3. For a discussion of the (non-Hippocratic) origins of this admonition, see Albert R. Jonsen, "Do No Harm: Axiom of Medical Ethics," in *Philosophical Medical Ethics: Its Nature and Significance*, edited by S. F. Spicker and H. T. Engelhardt, Jr., Dordrecht, Holland: D. Reidel Publishing Co., 1977.
4. Middleton, William W. "Ethical Process Enforcement and Sanctions – The Engineering and Physical Science Societies," delivered at the AAAS-IIT Workshop on Professional Societies and Professional Ethics, May 23, 1986.
5. Based on a scenario by Dan Dunn, Chris Minekime, and John Van Houten, MIT '93.
6. A clevis is a U-shaped metal piece with holes, through which a pin is run to attach a drawbar to a plow.
7. Thorpe, James F. and William H. Middendorf. 1979. *What Every Engineer Should Know About Products Liability*. New York and Basal: Marcel Dekker, Inc., p. 34. Martin Curd and Larry May in their booklet, *Professional Responsibility for Harmful Actions* (Dubuque Iowa: Kendall/Hunt, 1984), discuss retrospective responsibility (which, following the discussion in chapter one above, can be called "judge problems") and criteria for ethical and legal fault, including the hammer case.
8. Bucciarelli, Louis L. 1985. "Is Idiot Proof Safe Enough?" *Applied Philosophy*, 2(4):49–57. Reprinted in *Ethics and Risk Management in Engineering*, edited by Albert Flores, Landam, New York & London: University Press of America, pp. 201–209.
9. For a fuller discussion of this case, including complexities omitted here owing to space limitation, see Martin Curd and Larry May, *Responsibility for Harmful Actions*, Dubuque, Iowa: Kendall/Hunt Publishing Co., pp. 16–21; Paul Eddy, Elaine Potter, and Bruce Page, 1979, "Is the DC-10 a Lemon?" *New Republic*, June 9:7–9. For a discussion of the economic causes of design decisions that contributed to both DC-10 crashes (and other airline accidents), see John Newhouse, 1982, "A Reporter at Large: The Airlines Industry," *The New Yorker*, June 21:46–93 and Newhouse's book from which this piece was taken, *The Sporty Game*, New York: Alfred Knopf, Inc., 1982.

10. Missouri Board for Architects, Professional Engineers and Land Surveyors vs. Daniel M. Duncan, Jack D. Gillum, and G.C.E. International, Inc., before the Administrative Hearing Commission, State of Missouri, Case No. AR-84-0239, Statement of the Case, Findings of Fact, Conclusions of Law, and Decision rendered by Judge James B. Deutsch, November 14, 1985, pp. 54–63. Case No. AR-84-0239.

11. Ibid., pp. 423–425. See also Pfrang, Edward O. and Richard Marshall, "Collapse of the Kansas City Hyatt Regency Walkways," *Civil Engineering-ASCE*, July 1982, pp. 65–68. This article contains the official findings of the failure investigation conducted by the National Bureau of Standards, U.S. Department of Commerce.

12. Administrative Hearing Commission, State of Missouri, Case No. AR-84-0239, pp. 63–66.

13. Ibid., p. 384.

14. The synopsis given here is primarily derived from W. M. Kim Roddis, 1993, "Structural Failures and Engineering Ethics," *Journal of Structural Engineering ASCE* (May) and from the Hyatt Regency case in the case materials, *Engineering Ethics*, edited by R. W. Flumerfelt, C. E. Harris, Michael J. Rabins, and C. H. Samson (Texas A&M), final report to the NSF on Grant Number DIR-9012252.

15. This is specified in the "Delaney Clause" in Section 409 of the 1958 Federal Food, Drug and Cosmetic Act (FFDCA) of 1958. See Senate Report No. 85-2422, 85th Congress, 2nd Session (1958) and 21 United States Code, Section 348(c) (A) (1976). For a discussion of proposals to change the standard for food additives see the Congressional Research Service, Report for Congress "The Delaney Clause Effects on Pesticide Policy" by Donna U. Vogt, Analyst in Life Sciences, Science Policy Research Division, updated July 13, 1995, 95-514 SPR. It is available on the WWW at http://www.cnie.org/nle/pest-1.html.

16. Dowie, Mark. 1977. "Pinto Madness," *Mother Jones*, September/October: 19–32.

17. The provisions of the Violent Crime Control and Law Enforcement Act of 1994 (the "1994 Crime Bill") are summarized on the WWW at http://gopher.usdoj.gov/ crime/crime.html. Among the newspaper articles summarizing provisions of the act is John Aloysius Farrell's "US House OK's $30.2b crime bill," *Boston Globe*, Aug. 2, 1994, pp. 1, 8.

18. In 1988, 31% of all Medicare expenses occurred in the last year of patient lives. Since only 5% of those entitled to Medicare die in any one year, this is a disproportionately high expenditure. Forty percent of the final year total was expended for care in the last 30 days of a person's life. Although some of this care was expected to enable the patient to resume a meaningful life, as those of us who teach in hospitals know, some is futile and given primarily because the family, staff, or patient cannot accept with the reality of the situation. Although the final year total is a large proportion of the total health care budget, that proportion did not rise markedly from 1976 when the figure was 27%, to 1988. The absolute dollar amount quadrupled in that period, however, from $3,488 to $13,316 per Medicare recipient. See Knox (1993).

19. *Medical Ethics Advisor*, Jan. 1991, pp. 1–4.

20. Based on a scenario by Peter Kassakian, MIT '94.

21. Based on a scenario by Wayne Lam, MIT '95.

22. Rowe, Mary P. and Baker, Michael. 1984. "Are You Hearing Enough Employee Concerns?" *Harvard Business Review*, 62(3):127–135; Michael Davis, "Avoiding the Tragedy of Whistleblowing," *Business & Professional Ethics Journal* 8(4):3–19; Peter Block, *Stewardship – Choosing Service Over Self-Interest*, San Francisco: Berrett-Koehler Publishers, 1993; Elliston, Keenan, Lockhart, van Schaick, *Whistleblowing: Managing Dissent in the Workplace*, New York: Praeger Scientific, 1985.
23. This case, "Whistleblowing City Engineer," and the Board's discussion of it and two related cases appear in *Opinions of the Board of Ethical Review*, Volume VI, Alexandria, VA: National Society of Professional Engineers, 1989, pp. 115–116.
24. This case, "Public Criticism of Bridge Safety," and the Board's discussion of it and of three related cases appear in *Opinions of the Board of Ethical Review*, Volume VI, Alexandria, VA: National Society of Professional Engineers, 1989, pp. 117–119.

4

TWO MODELS OF PROFESSIONAL BEHAVIOR: ROGER BOISJOLY AND THE *CHALLENGER*, WILLIAM LEMESSURIER'S FIFTY-NINE STORY CRISIS

PART 1. ROGER BOISJOLY'S ATTEMPTS TO AVERT THE *CHALLENGER* DISASTER[www]

What do safety problems look like to the engineer who encounters them? What are good ways of responding to such problems at each stage of their development? Much can be learned from the attempts of Roger Boisjoly, an engineer at Morton Thiokol, to avert the *Challenger* disaster in January 1986. His care in coping with the uncertainties about the nature and extent of the threat to the shuttle and his courageous persistence in raising issues exemplify responsible engineering behavior.

Like others who have spent time with Roger Boisjoly I find him an exceptionally sincere and forthright person who is not only truthful with others but also honest with himself about his feelings and motives, a person not likely to fall into self-deception. His integrity and openness make his personal account of events especially illuminating.

MORAL LESSONS FROM ROGER BOISJOLY'S RESPONSE TO SAFETY PROBLEMS

In hindsight, ascribing blame for accidents and disasters based on the outcome is tempting. Any action that would have prevented the fatal *Challenger* flight, for instance, may seem justified; any failure to take an action to stop the flight may look like a mistake. But this view is superficial. Judging actions solely by their outcomes omits consideration of the other harmful consequences such actions risk and tells us nothing about how to act in situations in which we cannot perfectly foresee the outcome – the kind of situation in which we usually find ourselves. The challenge for responsible action is not merely to avoid one possible negative outcome, but to achieve a generally good outcome.

To take Roger Boisjoly's actions as exemplary does not mean that they are above criticism or that they could not in any way be improved on. Exemplary actions like excellent designs may be improved on, but they give us the shoulders of giants to stand on. Some have suggested, for example, that if Boisjoly had made a more effective graphical presentation of the data on hot gas blowby of previous shuttle flights, he could have made his case more convincing. If this is a valid criticism, it provides a lesson for engineering educators, since effective graphical presentation is a skill that receives little or no emphasis in most engineering programs.

The point of examining Boisjoly's experience is to learn from this example what engineers do when they respond well to moral problems such as safety problems. Evaluation of the strengths and weaknesses of the shuttle program has already been done in the report of the Presidential Commission [1] (commonly called the "Rogers' Commission" for William P. Rogers who chaired it) and in several books, including Malcolm Mc-Connell's *Challenger: A Major Malfunction.*

For his effort to avert the *Challenger* explosion, Boisjoly received the AAAS award for Scientific Freedom and Responsibility. The implication of the AAAS award is that Boisjoly's actions were well-conceived and of the right sort to bring attention to the risks he recognized to the *Challenger* flight. But acting well does not guarantee a good outcome, if one does not have complete control of the situation. Even the best practitioner rarely has control of all the factors that determine the outcome. As we shall see later in this chapter, William LeMessurier was successful in averting the collapse of the Citicorp Tower in New York City because he had more power to make his voice heard and because other figures also behaved responsibly. Most engineers will more often have only the level of control that Boisjoly had.

BACKGROUND AND THE POSTFLIGHT INSPECTION IN JANUARY 1985

For Roger Boisjoly the story began in January 1985, a year before the flight of the *Challenger.* At this time, Boisjoly was involved in the postflight hardware inspection of another shuttle flight, Flight 51C. [2] During this inspection he observed large amounts of blackened grease between the two O-ring seals, showing that the grease had been burned by escaping combustion gases. Gases from the rockets had, under immense pressure, created a blowhole through more than ten feet of zinc-chromate putty. Some hot gas had blown by the primary O-ring seal as well. Were the gas to leak by the secondary seal, it would risk igniting the fuel tanks, causing them to explode.

Boisjoly reported these findings to his superiors, who asked him to go to the Marshall Space Flight Center in Huntsville, Alabama to give a presentation of his observations and explain the seal erosion and escape of hot gases that caused it. Boisjoly said he believed that the resilience of the O-rings, and hence their capacity to seal, had been compromised by lower than usual launch temperatures.

BEING ASKED TO TONE DOWN THE HYPOTHESIS ABOUT COLD TEMPERATURE

NASA asked Morton Thiokol to give a more detailed presentation on the seal function as part of the flight readiness review for flight 51E, scheduled for April 1985. Boisjoly presented his views at three successively higher-level review boards, but NASA management insisted that he soften his interpretation for the final review board.

The primary seal in Flight 51C, the January 1985 flight, had leaked gas in what was the most extreme temperature change in Florida history. That such conditions would soon recur was unlikely, certainly not for Flight 51E which was scheduled for launch in Florida in April. Before pressing his hypothesis that low temperature had been a factor in the failure of the seals, Boisjoly took the opportunity to test his hypothesis. He sought out his friend and colleague Arnie Thompson to discuss the blowby and the effect of cold temperature on an O-ring's resilience.

BOISJOLY'S ACTION: DISCUSSING CONCERNS WITH PEERS

The situation in which Boisjoly found himself was not an emergency, and he had respected colleagues within his work group. In such circumstances, talking over the situation and using one's own interpretation of it are good ideas because they enable one to:

- check one's own perceptions and interpretations,
- develop peer support for one's concerns, and
- get practical suggestions and help in taking the next step in addressing those concerns.

Thompson proposed conducting tests of the effect of temperature on resiliency, which he and Boisjoly then carried out.

BOISJOLY'S ACTION: CONDUCTING TESTS

If the situation is not an emergency, it is a good idea to define the risk as precisely as possible before carrying concerns outside one's immediate work group. This may require conducting experiments to improve one's estimate of the risk or to test the hypothesized causes of the risks.

However, if the test involves large expenses in time or equipment, securing approval to conduct the tests may first require making a good case for the hazard you perceive.

The resiliency testing showed that low temperature was a problem. Boisjoly and Thompson discussed the data with Morton Thiokol engineering managers, but the managers felt the finding too sensitive to release.

STAGNATION IN THE FACE OF MOUNTING EVIDENCE ABOUT SEAL EROSION

Another postflight inspection occurred in June 1985 at Morton Thiokol in Utah. This time a nozzle joint from Flight 51B, which flew on April 29, 1985, was found to have a primary seal eroded in three places over a 1.3-inch length. Inspectors postulated that the primary seal had never sealed during the full two minutes of flight. The secondary seal in the same joint also showed signs of erosion. Boisjoly was greatly concerned about this finding.

A flight readiness review presentation was prepared for Flight 51F, scheduled for launch on July 29, 1985. The status of the booster seals was the topic of a presentation to NASA at Marshall Space Flight Center on July 1, 1985 and of another the next day. The preliminary results of the O-ring resilience testing in March were presented for the first time during this meeting. Everyone in the program was by then aware of the influence of low temperature on the joint seals.

At this point, management at Morton Thiokol and NASA had evidence that the seals did not perform as required and that cold temperature was a factor in their failure to perform. An attempt on July 19, 1985 to form a team to work on the seal erosion problem failed, however. In his journal, Boisjoly recorded his frustration with management's failure to take appropriate steps in response to the persistent failure of the O-rings.

BOISJOLY'S ACTION: KEEPING A JOURNAL

By this time Roger Boisjoly's heightening concern led him to begin keeping a journal of events pertaining to seal erosion. The journal would later become an important aid to him in giving testimony to the Presidential Commission investigating the *Challenger* explosion (the "Rogers Commission"). The journal had the immediate purpose of enabling Boisjoly to monitor events so that he could discover and remove roadblocks that stood in the way of fixing the problem.

Roger Boisjoly reports being influenced by the memory of an engineer who had been involved in another famous disaster: the 1974 DC-10 crash. For many years this crash was the worst airline accident in terms of loss of life. The engineer whose designs had significantly contributed to this accident became almost totally dysfunctional for a long period after the crash and went through his workday under heavy sedation. This memory made Roger Boisjoly realize that major safety problems might not be remedied through normal reviews. Therefore, he closely monitored all action taken on the problem of the O-ring seals.

The problem had evolved from a matter of seal erosion during several flights to one experimentally shown to be aggravated by cold temperature. Although no launches in cold temperature would occur in midsummer, the steps necessary to assure adequate sealing in all weather were not being taken. Roger Boisjoly, therefore, took direct action to force the attention of management to the issue. He wrote a memo to the vice president of Engineering, Robert Lund, stating his concern that failure to address the problem would mean an explosion of the shuttle. This memo was immediately stamped "company private," meaning that it must not be sent outside Morton Thiokol.

This memo was specifically cited in the AAAS award to Boisjoly for his efforts to avert the *Challenger* explosion. The persistent inability of management to take effective action on a major threat to safety justifies and even requires speaking up, even if it requires going outside normal channels.

BOISJOLY'S ACTION: EXPRESSING A CONCERN

Although the situation that Roger Boisjoly faced at this point was not an emergency, since no low-temperature launch was imminent, Boisjoly had good reason to think that management was not taking steps to properly address the hazard evidenced in seal erosion. He chose to communicate his concerns to Vice President Bob Lund in writing, although on other occasions Boisjoly had expressed his concerns by talking with Lund. Written communication ensures that a decision maker has a precise statement of the problem and that anyone else who is shown or who receives a copy of the communication sees the same statement, which they can each later review.

Some corporations, such as automobile manufacturers who have experienced lawsuits charging unsafe design in which employee memos have been subpoenaed, are reluctant to create records that might be used against them in some future liability suits. They may discourage hard copy (paper and ink) communication about safety problems. To retain the precision of written communication, some of these companies encourage the use of electronic mail. However, e-mail is rapidly acquiring the same potential as evidence in court.

BOISJOLY'S ACTION: INFORMING OTHERS WHEN GOING TO THE TOP

Inexperienced professionals who find that they are in a situation in which they need to go "over someone's head" or "to the top" with their concerns may neglect the question of whom they should inform or consult in doing so. Going to the top is less likely to offend those who are "leapt over" if they are at least informed, so that they are not caught unprepared for the actions that follow. Following his usual practice, Boisjoly showed his memo to his direct supervisor, who then countersigned it.

Boisjoly's memo got the attention of top management, which then authorized the formation of a seal team.

A COMPANY'S CONCERN ABOUT ITS IMAGE

On August 19, 1985 Morton Thiokol personnel went to a meeting at the Marshall Space Flight Center on the problem with the seals on the booster rockets. In September 1985, NASA officials instructed Morton Thiokol to send a representative to the Society of Automotive Engineers (SAE) conference in October to discuss the seals and solicit help from others at the conference. Boisjoly was selected to make the presentation, but NASA gave strict instructions that he was not to express the critical urgency of the joint problem, but rather to emphasize the progress on solving it that the company had made thus far.

Every organization has difficulties that it overcomes in the normal course of its operation. A company is usually reluctant to publicize its mistakes and failures. At this time, NASA was receiving public criticism for promising commercial uses of shuttle flights that never materialized, for having cost overruns, and for being behind schedule, and Morton Thiokol was worried about losing its position as sole contractor for certain parts of the space shuttle effort.

In other cases, a company, a research lab, or an academic researcher might be reluctant to release information about mistakes that competitors might use to advantage. A major hazard in concealing difficulties is that the difficulties remain inadequately addressed. NASA's unwillingness to reveal a major malfunction in the solid rocket booster hampered Boisjoly's attempt to get expert advice about the seal problem. However, NASA's instruction was not so outlandish that Boisjoly felt morally obliged to defy it.

After his presentation, Boisjoly asked the audience for suggestions to improve the design, but he received none. Boisjoly and another Morton Thiokol engineer, Bob Eberling, spent the remainder of the convention meeting with seal vendors whom they had previously contacted for help.

WORKING WITH POOR MANAGEMENT SUPPORT

Although the seal task team had been formed in response to his July memo, Boisjoly reports that management did not give the effort much support and that it lacked necessary resources and information. Many unanswered questions remained as the seal task team approached the end of 1985: Almost twenty flights had flown successfully and some of the cases of hot gas blowby had occurred during warm as well as cold temperatures.

In attempting to bring attention to the problem, Boisjoly used normal channels to the fullest and had already presented his concerns directly to the vice president of Engineering. He continued to keep his journal of progress, or lack of progress, on the seal problem and used activity reports to document the frustration of his efforts, including attempts to get more data on seal erosion. He received no response, however, and never knew if his comments went to upper management. The Presidential Commission investigating the shuttle disaster later cited frequent failures to pass along vital information as a principal pattern of errors that led to the shuttle disaster.

Contacting an ombudsman or a safety hotline at Morton Thiokol or at NASA might have been appropriate at this point, but Morton Thiokol and NASA had neither.

THE DAY AND EVENING BEFORE THE FLIGHT

The *Challenger*, with a crew that included "the teacher in space," was scheduled to fly on January 28, 1986. The preceding day, Boisjoly and his colleagues were shocked to learn that the overnight temperature at the launch site was predicted to be only 18 degrees Fahrenheit, lower than the record cold experienced the previous year. Boisjoly and several of his colleagues were firmly convinced that this extreme weather condition presented a major threat to the capacity of the O-ring seals to function and thus to the survival of the flight crew.

With time running out, Boisjoly and his colleagues went directly to Bob Lund, the vice president of Engineering, to make their case for postponing the flight. They convinced him of the danger and secured his decision to recommend against flying. To make their point to NASA at the teleconference scheduled for that evening, they hurriedly prepared viewgraphs outlining their concerns about launching at such a low temperature.

The teleconference linked Morton Thiokol with Kennedy Space Center (KSC) in Florida and the Marshall Space Flight Center (MSFC) in Huntsville, Alabama. A manager colleague who had long shared Boisjoly's concerns, Al McDonald, was present at KSC for the teleconference ("telecon"). Discussion started with a history of O-ring erosion in field joints. Boisjoly reports that

> data was presented showing a major concern with seal resiliency and the
> change to the sealing timing function and the criticality of this on the

ability to seal. I was asked several times during my portion of the presentation to quantify my concerns but I said I could not since the only data I had was what I had presented and that I had been trying to get more data since last October.

When Boisjoly made this last comment, the general manager of Morton Thiokol glared at him.

This presentation ended with the recommendation not to launch below 53 degrees. NASA then asked Joe Kilminster, vice president of Space Booster Programs at Morton Thiokol, for his launch decision. Kilminster said that because of the engineering judgment just presented he would recommend against launching. Then Larry Mulloy of NASA at KSC asked George Hardy of NASA at MSFC for his launch decision. George responded that he was appalled at Thiokol's recommendation against flying, but said he would not launch if Morton Thiokol objected. Mulloy then spent some time giving his interpretation of the data, arguing that the data were inconclusive.

The vehement reaction of NASA's George Hardy to the recommendation against launch surprised Boisjoly. Not only was Hardy usually moderate in speech, but in Boisjoly's experience, NASA had shown a great concern for safety consciousness. Nevertheless, the Rogers Commission was to find that, although failure of the seals on the field joints caused the explosion, many other flaws in the shuttle design and poor patterns of communication might also have resulted in a fatal crash. NASA's prior reputation for safety seems to have rested on their practice of placing the burden of proof on those who advocated launch. If there was any question of a risk, a flight was normally postponed. This time, however, NASA did not follow its established practice.

Joe Kilminster responded by asking for a five-minute off-line caucus to reevaluate the data. The mute button was pushed, so the two NASA groups could no longer hear Morton Thiokol's discussion. Immediately Thiokol's general manager, Jerry Mason, said in a soft voice: "We have to make a management decision." It would be a mistake to interpret Mason's remark to mean that he thought that from management's point of view the explosion of the *Challenger* would be an acceptable outcome. Mason intended to consider factors other than safety, factors that would weigh in favor of launching. To consider other factors meant downplaying the danger that Boisjoly knew to exist. Boisjoly reports that he became furious when he heard Mason's remark. He describes the subsequent discussion as follows:

Some discussion had started between the managers when Arnie Thompson moved from his position down the table to a position in front of the managers and once again tried to explain our position by sketching the joint and discussing the problem with the seals at low temperature.

Arnie stopped when he saw the unfriendly look in Mason's eyes and also realized that no one was listening to him. I then grabbed the photographic evidence showing the hot gas blowby and placed it on the table and, somewhat angered, admonished them to look and not ignore what the photos were telling us. I, too, received the same cold stares as Arnie with looks as if to say, "Go away and don't bother us with the facts."

At that moment I felt totally helpless and that further argument was fruitless, so I, too, stopped pressing my case... .

During the closed managers' discussion, Jerry Mason asked in a low voice if he was the only one who wanted to fly. The discussion continued, then Mason turned to Bob Lund, the vice president of Engineering, and told him to take off his engineering hat and put on his management hat. The decision to launch resulted from the yes vote of only the four senior executives since the rest of us were excluded from both the final decision and the vote poll. The telecon resumed and Joe Kilminster read the launch support rationale from a handwritten list and recommended that the launch proceed.

NASA promptly accepted the recommendation to launch without any probing discussion and asked Joe to send a signed copy of the chart.

The change in decision so upset me that I do not remember Stanley Reinhartz of NASA asking if anyone had anything else to say over the telecon. The telecon was then disconnected so I immediately left the room feeling badly defeated.

In this situation, NASA responded in a way that was unprecedented in Boisjoly's experience. In the wake of the success of the *Apollo* project that had placed astronauts on the Moon, NASA had conceived the shuttle project to retain public support. They had overpromised achievement. Having failed in many of its promises, NASA felt pressure to have the flight with "the teacher in space" as a visible success that could be mentioned in the State of the Union Address, which the President was to give the next day. As Boisjoly was later to find out, Morton Thiokol was at this point negotiating a new contract with NASA and trying to remain the sole contractor for the Solid Rocket Booster Program. This negotiation undoubtedly influenced top management's decision not to delay the *Challenger* flight, and so they reversed their engineering decision.

Faced with a horrible outcome, it is tempting to wish for a miraculous "rescue." Thus it is often suggested that Boisjoly, Thompson, or

McDonald ought to have done something more to see that the flight was stopped. Calling the press or notifying the astronauts are favorite suggestions. But calling the press would have been excessive before the teleconference since Bob Lund had already agreed to postpone the flight. After the teleconference, a media story would have come too late, even if Roger Boisjoly had known a responsible journalist whom he could have readily contacted.[3]

The astronauts themselves are sequestered the night before each flight. Had someone called their families, it is not clear how they would have interpreted a call from a stranger claiming to be a Morton Thiokol engineer who thought the flight was unsafe, what they could have done with such information, or whom they could have convinced. Those like Boisjoly and Thompson who took many personal risks to bring their safety concerns to the fore have regrets about the outcome but few about their own behavior. Boisjoly says he only regrets not vigorously advocating for a change in the temperature criteria for launch to fifty degrees or more when he first became aware of the cold temperature threat. Bob Eberling, one of the engineers who agreed with Boisjoly and Thompson but did not confront management the night of the teleconference, has said, in January of 1996 on the television program "Sixty Minutes," that he regrets he did not do more that night.

The Presidential Commission found multiple instances in which safety concerns were not passed along. The communications about safety problems were found to be so grave that astronaut Sally Ride, who served on the Commission, refused to fly any more after the Commission's hearings. As Commission member Richard Feynman put it, "The guys at the top ...didn't want to hear about the difficulties of the engineers... It's better if they *don't* hear so they can be more 'honest' when trying to get Congress to OK their projects."[4] NASA's deputy administrator, Hans Mark, wrote in an article in the *IEEE Spectrum* that he had known and written a memo about the O-ring problem two years before, so that it seemed incredible to him that Jesse Moore, the NASA decisionmaker in charge of the *Challenger* flight, did not know of the O-ring problem. Thus, even if Boisjoly *had* been able to find someone in authority at NASA to tell that evening, he would have been giving them information that they were already showing resistance to receiving.

Later, Boisjoly did take information to outside authorities, thereby becoming "a whistle-blower." He gave documents to the Presidential Commission investigating the *Challenger* explosion without first giving them to Morton Thiokol to review and censor. His disclosure of the information to the Presidential Commission led to sanctions against him at Morton Thiokol and to his being ostracized in the little Utah town where he had previously been mayor. (The story of Boisjoly's actions after the *Challenger* disaster may be found in the WWW Ethics Center for Engineering and Science.) Given his willingness to give documents to the Presidential Commission despite the risk to his career and his continuing to reside in a town for which Morton Thiokol was the principal employer, it is reasonable to assume that had he known of an effective way of raising his concern about the effect of temperature on the seals, he would have done so. To require that professionals behave responsibly cannot imply that they should be blamed for bad outcomes or that others are not responsible for supporting their efforts to safeguard the public's safety, health, education, or well-being.

In this instance those in authority behaved not only unreasonably, but also unpredictably. An engineer could not have known in advance that going outside the company was necessary. If they are prepared to raise safety issues as clearly, forthrightly, and persistently as Boisjoly and some of his colleagues, then engineers will be meeting their responsibility for safety – although they may not be able to prevent every disaster. If corporations and government agencies support engineers who raise safety concerns, a single bad decision will not create a disaster like the explosion of the *Challenger*.

Accidents and safety problems are clearly not in the interest of management. When the general manager of Morton Thiokol recommended launching over the objections of his engineers, he made a very bad management decision.

PREVENTING ACCIDENTS

The case histories of technological accidents and related health hazards reveal few instances of flagrant disregard and suppression of evidence on health and safety risks. The behavior demonstrated by the asbestos industry and by A. H. Robbins, makers of the Dalkon Shield, are the exception. Fewer companies even undertake the cynical comparison of the cost of legal liabilities for death and injury to the cost of preventing accidents (as Ford did with the Pinto gas tank). Cases such as those of the Dalkon Shield and the Pinto gas tank have received the greatest press attention precisely because of the outrageousness of the decisions involved. A review of case histories of accidents shows that the *Challenger* case is more typical of poor management decisions that result in accidents: Usually management does not flagrantly disregard safety, but management fails to give due regard to safety hazards because of the pressure of deadlines or financial exigencies.

Using data comparable to that Boisjoly was able to gather, participants in an exercise now widely used in management training often behave as Morton Thiokol managers did. The trainees are asked to make a decision as individuals and then as members of a group. As individuals, the majority decide not to take the life-threatening risk in the face of the evidence. In groups, however, two thirds of the groups in predominantly male training programs decide to take the risk. (Most of the groups in an all-female management program did not.)[5]

A review of cases of accidents fails entirely to pick up the many situations in which managers and corporations responded appropriately to the safety concerns of their engineers. Many corporations recognize that it is in their interest to provide their employees with adequate opportunities to raise concerns about safety, and they are taking steps to provide their employees with avenues to raise concerns about safety and other ethical matters. Attention to how organizations support the timely expression of employee's ethical concerns is at least as important to preparing for professional responsibility as reviewing disasters. Therefore, this book discusses good ways that universities, departments, companies, and organizations foster responsible action by their members. This information is meant to help readers to assess the ethical climate of an organization before they join it.

NOTE ON THE *CHALLENGER* DISASTER AS A FORMATIVE
EXPERIENCE FOR MANY ENGINEERS AND FOR
POPULAR CULTURE

For the present generation the *Challenger* accident functions as a so-called flash-bulb memory, much as the memory of the Kennedy assassination did for the previous generation. Many young people today can remember where they were when they heard about, or saw, the *Challenger* explosion. Often it was watching the launch with "the teacher in space." Many testify that when the explosion occurred, their teachers (perhaps because they were so shocked) were unable to talk about what had occurred and simply ushered them back to their classes. Some students were lucky enough to have teachers or parents who could talk to them about what had occurred, but a good number of the older students recall it as an experience of confusion or disillusionment with authority figures.

The case has provided not only another famous accident for study, but also a personal experience of people being reluctant to deal forthrightly with bad news. This formative experience for engineers born between the mid-1950s and the mid-1970s is likely to affect the culture of engineering for years to come.

Roger Boisjoly's attempt to avert the *Challenger* disaster deserves careful study for what it reveals about fulfilling an engineer's professional responsibility for safety. Boisjoly's problem situation and the responses he made to it at various stages of development illustrate many points about how safety problems are recognized and the actions needed to resolve those problems.[6]

PART 2. WILLIAM LEMESSURIER'S HANDLING OF THE "FIFTY-NINE STORY CRISIS" [7www]

The Citicorp Tower in New York City was completed in 1977. William J. LeMessurier designed the supporting structure for this unusual skyscraper. Shortly after its completion, LeMessurier discovered that the building was vulnerable to hurricanes: One of the strength that hits New York City about every sixteen years could bring down the building. Stresses caused by factors that, in 1977, were not normally considered in designing buildings, turned out to be unexpectedly high in such conditions. In addition, without LeMessurier's knowledge, a change had been made in the construction: Bolts had been substituted for welds, and too few bolts were used to handle even the stresses that LeMessurier had anticipated. The harrowing story of LeMessurier's discovery of the defect and correcting the flaw in his own design is another instructive example of a person fulfilling the moral responsibilities that go with being an engineer. Only recently has it been made public.[8] LeMessurier's timely and candid revelation of the danger, coupled with appropriate action of other key decisionmakers, averted what would have been a disaster of astounding proportions.

Much has been learned from the technical details of this near accident (LeMessurier drew on that experience when he later consulted on the equally serious problems with I. M. Pei's John Hancock Tower in Boston), and there is much to be learned from how LeMessurier went about eliminating the threat.

When Citibank began planning for a new headquarters tower in midtown New York, the art of designing and building a structurally safe skyscraper seemed nearly perfected. After the development of steel-frame buildings and Elisha Otis's successful introduction of the safety-brake-equipped elevator in the 1850s, architects began to design ever taller buildings. The Home Insurance Building constructed in Chicago in 1885 was the first multistoried building to have a complete structural frame supporting its masonry walls.

By the 1930s, when the 102-story Empire State Building was completed, skyscrapers had begun to appear in cities all over the world. Creative architects and engineers introduced further innovations in the design and construction of tall structures that called for lighter materials and columnar supports. Chicago's Hancock Building, for instance, incorporated an innovative system of diagonal bracing that allowed the building to be much leaner and lighter than would have been possible with the more customary structural supports.

LEMESSURIER'S INNOVATIVE DESIGN FOR THE CITICORP TOWER

William LeMessurier had already distinguished himself as a preeminent structural engineer by the time his structural engineering firm was engaged as consultant on a new corporate headquarters for Citibank. He had had extensive experience with skyscrapers. In the first building he designed, Boston's State Street Bank, he incorporated an inventive cantilever girder system. In the design of another of his buildings, the famous Boston Federal Reserve Bank, he designed an opening at the base of the building large enough for an airplane to fly through.

The Citibank headquarters needed LeMessurier's experience with innovative designs because of special design constraints at the site. A church congregation owned part of the block where Citicorp planned to build. Citicorp agreed to replace the old church building with a new free-standing structure, and in return gained the air rights above the new building. To allow for the church underneath, the Citicorp tower was constructed on nine-story stilts. Because the church was to be located at a corner of the site, rather than the middle, the stilts had to be under the middle of each of Citicorp tower's outer walls, rather than under its corners. This posed a challenging structural engineering problem. LeMessurier's imaginative solution was to use large diagonal girders throughout the building; these would transfer the tower's great weight to four huge columns that would run the height of the building on each side and anchor the structure to the ground. The new church could then be constructed as planned, underneath one of the tower's corners.

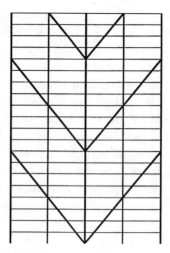

Figure 4.1. Diagonal Girders (Twenty Stories Shown).

LeMessurier's innovative diagonal bracing for the Citicorp tower greatly reduced the weight of the structure. The reduced density of the tower also makes it more dynamically excitable: It would have had a disturbing tendency to sway in the wind had it not been for a tuned mass damper. This damper, installed at the top of the building, consists of a 400-ton concrete block floating on pressurized oil bearings attached to two horizontal springs at right angles to one another. The innovation of the damper made this the first building to have a mechanical aid as part of its design.

THE DISCOVERY OF THE CHANGE FROM WELDS TO BOLTS

In May 1978, LeMessurier, acting as structural consultant to another building being planned in Pittsburgh, again thought of using a sort of diagonal brace as part of his design. As in the Citicorp tower, his plans called for full-penetration welds to hold the sections of the braces together. Welded joints are extremely strong, but as a potential contractor for the Pittsburgh construction job pointed out, they are time consuming to make and much more expensive than bolted connections. At this point, LeMessurier learned that the substitution of bolts for welds had also come up during the Citicorp tower's construction, and the Citicorp contractors had decided to put the braces together with bolted joints to save the cost of welding. The New York contractors had agreed that bolts would be strong enough to make the structure safe.

When LeMessurier referred the Pittsburgh contractor who was concerned about the cost of welding to the successful Citicorp job, he learned of the substitution of bolts for welds. Initially LeMessurier was not alarmed, since the substitution was reasonable from an engineering standpoint. Since LeMessurier had served merely as a consultant he saw no reason to have been previously informed. His assessment of the effect of the substitution was to change, however, as he discovered evidence of another danger compounded by the substitution.

INVESTIGATING THE EFFECTS OF QUARTERING WINDS

In June 1978, a month after LeMessurier was told of the switch from welds to bolts in the Citicorp building, he received a telephone call from a student. This student's professor had been studying LeMessurier's design for the Citicorp tower and had concluded that LeMessurier had put the building's nine-story supports in the wrong place. They belonged on the tower's corners, according to this professor, not at the tower's midpoints.

The professor had misunderstood the design problem that had confronted LeMessurier, so LeMessurier explained to the student his line of reasoning for putting the tower's supports at the building's midpoints. He added that his unique design, including the supports and the diagonal-brace system, made the building particularly resistant to quartering, or diagonal, winds – that is, winds coming in on the diagonal and thus hitting two sides of the building simultaneously.

Shortly thereafter, LeMessurier decided that the subject of the Citicorp tower and quartering winds would make an interesting topic for the structural engineering class he taught at Harvard. Since at the time the requirements of the New York building code (or of any building code) extended only to the effects of perpendicular winds, LeMessurier had not calculated the effects of quartering winds.

When LeMessurier did the computations, he found that for a given quartering wind, stresses in half of a certain number of structural members would increase by 40%.

Then he became concerned about his welds being replaced by bolts. Had the New York contractors taken quartering winds into account when they replaced the welds with bolts? Had they used enough bolts? The second question was particularly important: A 40% increase in stress on certain structural members resulted in a 160% increase of stress on the building's joints, so it was vital that enough bolts be used to ensure that each joint was the proper strength.

What he found out was disturbing. The New York firm had disregarded quartering winds when they substituted bolted joints for welded ones. Furthermore, the contractors had interpreted the New York building code in such a way as to exempt many of the tower's diagonal braces from loadbearing calculations, so they had used far too few bolts.

Shaken, LeMessurier reviewed old wind-tunnel tests of the building's design, tests that had modeled a large part of midtown Manhattan. When he compared these against his new quartering-wind calculations, he discovered that under adverse weather conditions, the tower's bracing system would be put under even further stress.

WIND TUNNEL EVIDENCE OF THE DANGER

LeMessurier now believed there might be grave danger. He turned to Alan Davenport, a Canadian consultant during the building's design phases, for further confirmation. Davenport, who had run the original wind tunnel tests, now ran the tests again, with new calculations to reflect quartering winds and the change from welds to bolts.

The results, when compared with the building's original testing, confirmed LeMessurier's suspicion that stress in some of the building's structural members would increase. His concern grew, for the results indicated that a 40% theoretical increase in a member's structural stress would be much greater under real-world conditions. During a storm, the whole building could shake, causing all the structural members to vibrate synchronously.

LeMessurier worked through the revised wind tunnel data and quickly discovered that the building was vulnerable to a total structural failure: If a storm pulled a joint apart on the thirtieth floor, the whole building would collapse. A "sixteen-year storm" would have the strength to cause total structural failure if the storm knocked out electrical power necessary for running the building's damper. The tuned mass damper had an enormous steadying effect on the building and might help to reduce the stress on that joint.

The engineering problem was easy to solve; heavy steel-welded "band-aids" over the joints would give the building more strength than it was ever designed to have. The repairs needed to be finished before a hurricane hit, however. It was then the last day of July, and hurricane season was just beginning. To accomplish the repair, LeMessurier would have to announce the building's vulnerability and take responsibility upon himself. Doing so could cost him his career and reputation as a structural engineer. He could not predict the reception of his news by Citibank leadership, city officials, or the general public.

MOBILIZING SUPPORT

On July 31, LeMessurier contacted the lawyer of the architectural firm that had retained him as its structural consultant for the Citicorp tower and then the firm's insurance company. As a result, a meeting was arranged for the following day with several lawyers for the insurers, to whom LeMessurier related the entire story. The lawyers soon decided to bring in a special consultant, Les Robertson, a respected structural engineer. Robertson listened to LeMessurier's description of the situation and then took a more pessimistic view than even LeMessurier. Robertson did not believe that, for instance, the tuned mass damper would serve as a safety device despite LeMessurier's assurances that generators could keep the dampers running during an electrical power loss.

Citicorp had to be informed of the danger, so LeMessurier and his partner tried to contact Citicorp's chairman, Walter Wriston. Initially, Wriston was not available to them, but LeMessurier's partner was able to arrange a meeting with Citicorp's executive vice president, John Reed, who had engineering experience and who had played a part in the construction of the tower. Once more, LeMessurier detailed the situation. When prompted for a cost estimate, he guessed that one million dollars would be sufficient. He also explained that the repairs could be done without inconvenience to the tenants by isolating the bolted joints within plywood "houses" and doing the necessary work at night within those houses.

Reed appreciated the gravity of the situation and arranged a meeting with Walter Wriston on August 2. LeMessurier again told his story. Much to his relief, Wriston recognized the importance of the tower as Citicorp's new corporate emblem, and so he readily agreed to the repair proposal. He approved a plan to install emergency generators as a backup power supply for the tuned mass dampers and oversaw much of the relations with the public as well as with the building tenants.

The next day, LeMessurier met with two engineers from the construction company that was to perform the repairs. After examining the joints these engineers approved LeMessurier's plan to reinforce the bolted joints with welded bandaids.

Before undertaking the repairs, several steps were necessary. LeMessurier contacted the company that had constructed the tuned mass damper to help ensure the device's continuous operation. Meteo-rological experts were retained to give advance warning of any storm that could cause the building's destruction. LeMessurier reluctantly agreed with Robertson that, as a further precaution, an emergency evacuation plan for the building and the ten-block-diameter surrounding neigh-borhood be drafted. In its final form the plan was to involve up to 2,000 emergency workers provided by the Red Cross.

LeMessurier had to explain the situation to city officials both to se-cure their cooperation with the evacuation plan and to comply with the building code. They responded with approval and encouragement, rather than with the cynicism that LeMessurier expected. They too rec-ognized both the seriousness of the problem and the immediate need to solve it. Energy was not wasted on rancor or placing blame.

The final task and the one LeMessurier most dreaded was informing the press of what was going to be a major undertaking on the brand-new Citicorp tower. An initial press release was issued, indicating that the building was being refitted to withstand slightly higher winds. This was true to some extent, for the meteorological data suggested that the winds for that year were going to be somewhat higher than normal. But the *New York Times*, for one, was sure to express further interest in what could be a very juicy story. After an initial phone call from a reporter, LeMessurier found an unexpected reprieve in a citywide press strike.

ACCOMPLISHING THE REPAIR WITHOUT CAUSING PANIC

Repairs to the Citicorp building commenced immediately. The plan of action was to expose each bolted joint in the building by ripping away the flooring and walls around it, to cover each joint with a plywood house to minimize any visible signs that things were awry with the building's structure, and to complete the repair welding at night when the tenants were not in the building, so as not to inconvenience them.

The pace of work was fast. Parts of the interior around the bolted joints were torn up at night and put back together in the morning. LeMessurier occupied himself with repair process calculations. Les Robertson calculated how to repair the joints and, suspecting that other components of the building could be vulnerable, investigated the floors, columns, and braces for weakness.

The repair work was in full swing on the first of September, when a hurricane moving toward New York was detected. The news was met with alarm. Although the partial repairs, along with the tuned mass damper, greatly improved the building strength, no one wanted to see them tested. There was great relief when the hurricane moved out over the ocean.

Two weeks later, repairs had progressed to the point that, with no storms predicted, the elaborate evacuation plans could be scrapped. Repairs were completed the next month. Even if the tuned mass damper were to fail, a seven-hundred-year storm would not pose a threat to the Citicorp Center.

The engineering problem had been solved, and the building now exceeds even its originally intended safety factor.

THE FINAL TOUCH: LEMESSURIER'S GOOD NAME

LeMessurier feared for his career but did not allow any worries or self-protective impulses to sidetrack his attention from carrying out the repairs. In the middle of September, when work was almost complete, Citicorp notified LeMessurier and his partner that it expected to be reimbursed for the cost of the repairs.

The estimated total cost for the building's repair ranged from a high of $8 million for the structural work alone, given by one of the construction companies involved, to $4 million, which, according to LeMessurier, was the Citicorp estimate. (Citicorp did not make public its estimate of the cost of the repairs.)

LeMessurier's liability insurance company had agreed to pay $2 million, and LeMessurier brought that figure to the negotiating table. The Citicorp officials eventually agreed to accept the $2 million, to find no fault with LeMessurier's firm, and to close the matter.

A relieved LeMessurier nevertheless expected his insurance company to raise the premiums on his liability insurance. He would, he reasoned, appear as an engineer who had bungled an expensive job and caused the insurer to pay a large cash settlement.

At a meeting with officials from the insurance company, LeMessurier's secretary was able to convince them that LeMessurier had "prevented one of the worst insurance disasters of all time!"[8] Far from behaving in an incompetent or devious manner, LeMessurier had acted in a commendable way: He had discovered an unforeseen problem, acted immediately, appropriately, and efficiently to solve it, and solved it. LeMessurier's handling of the Citicorp situation increased his reputation as an exceptionally competent, forthright structural engineer. It also prompted his liability insurers to lower his premium.

PART 3. CONCLUSION: COMPARISON OF BOISJOLY
WITH LEMESSURIER

It is emotionally satisfying to close with a happy ending, but life is not always so obliging, as we saw in the case of the *Challenger* flight. Despite the dramatically different outcomes of the two cases, the responses of Boisjoly and LeMessurier to the problems they faced have much in common. The principal differences between the two were their positions, their ability to influence the outcome, and the intelligence with which their initiatives were received.

Both Boisjoly and LeMessurier were alert to the first evidence of danger. Both sought further information from testing and from the advice of others. Boisjoly worked with peers and was embedded in the organizational structures of Morton Thiokol and NASA at every stage of his efforts. He enlisted the support of those who would listen and made appropriate use of the organizational channels open to him. Although more solitary in the initial discovery of a threat, LeMessurier also showed a similar resolve to pursue the difficult safety questions that faced him.

The poor response of Morton Thiokol and of NASA to early evidence contrasts strongly with the cooperation that LeMessurier received from Citicorp and from city officials. Boisjoly overcame the resistance he faced by taking the highly unusual measure of writing directly to Morton Thiokol's vice president of Engineering. Both Boisjoly and LeMessurier took appropriate actions that they had never witnessed anyone else take. Despite the profound difference in the ultimate outcomes, both Boisjoly and LeMessurier demonstrated how courage, honesty, and concern for safety are implemented in engineering practice.

NOTES

[www]A hypermedia account of Roger Boisjoly's efforts in the year leading up to the *Challenger* flight is available at the WWW Ethics Center for Engineering and Science at http://ethics.cwru.edu in the section on moral leaders.

1. A link to the www version of this report may be found from the story of the *Challenger* in the WWW Ethics Center for Engineering and Science.
2. NASA had found severe erosion of O-ring seals from some flights before 1985, the starting point of Boisjoly's account.
3. For a perspective from within the engineering profession on the appropriate use of the media in raising ethical concerns, see the NSPE Board of Engineering Review discussion of Case 88-7. This case, "Public Criticism of Bridge Safety," and the Board's discussion of it and three related cases are available in the WWW Ethics Center for Engineering and Science at http://ethics.cwrd.edu. They originally

appeared in *Opinions of the Board of Ethical Review,* Volume VI, Alexandria VA: National Society of Professional Engineers, 1989, pp. 117–119.

4. Unger, Stephen H. 1994. *Controlling Technology: Ethics and the Responsible Engineer,* second edition, New York: John Wiley & Sons, Inc.

5. Daniel Goleman's *Vital Lies, Simple Truths,* (New York: Simon & Schuster, Inc., 1985), and Irving Janis's *Groupthink,* 2nd edition, (Boston: Houghton Mifflin, 1982) present theories about why people are reluctant to bring forward unwelcome information.

   Using the same data and individual and group decision format, but a more disguised story line with upper-level *engineering* students at several universities, I have gotten a different result. In my experience, if one of the students points out how the pattern in the data gives clear evidence of a hazard, the group decides not to take the risk. Some of the same engineering students go on to management programs whose members showed the group behavior of displacement toward risk taking, suggesting that norms for group behavior or the value of looking like a risk-taker may be different in engineering programs from those in (predominantly male) management programs.

6. The description of these attempts given here draws heavily on the account that Roger Boisjoly gave in January 1987 in an address at MIT, on another address that he gave in September of 1989, and on answers to questions he gave to those audiences and subsequently. The text of the first talk is published in several places, e.g., *Books and Religion,* Vol. 15, Nos. 3–4, March/April 1987, pp. 3–4, and Jonson, 1991, pp. 6–14.

   A videotape of this talk with questions bringing out the issue of responsibility discussed here is available from Roger M. Boisjoly, P. E., P.O. Box 248, Roosevelt, Utah 84066-0248. Boisjoly's address was part of a student-initiated course given in Independent Activities Period in January 1987. Robin Wagner, a graduate student in the Technology and Policy Program, led in organizing this course and took the initiative of inviting Roger Boisjoly to speak.

7. The story of William LeMessurier's resolution of the crisis with the Citicorp Tower appears in the WWW Ethics Center for Engineering and Science. Included in the site are slides showing Chicago's Hancock Building and other innovative skyscrapers as well as the Citicorp Tower. These are from an address LeMessurier gave in 1995. A videotape of this talk is available through the WWW Ethics Center.

8. It is told in the May 29, 1995 issue of *The New Yorker,* pp. 45–53.

9. *Ibid.,* pp. 45–53.

5

WORKPLACE RIGHTS AND
RESPONSIBILITIES

As we saw in Chapter 3, it is in everyone's interest that engineers be heeded when they foresee risks and threats to the public welfare. It is in a company's interest to see that engineers' concerns are heard within the company, rather than only after they have gone outside to "blow the whistle."[1] In prior chapters we have focused on the moral skills that enable engineers to fulfill their responsibilities both in responsive and unresponsive organizations. In the United States and other countries where employee engineers usually have no written employment contracts (as they do in Germany for instance[2]), companies may retaliate against engineers for pursuing ethical concerns that clash with their company's short-term business objectives. Therefore, creating a workplace that is relatively free of the risk of such harassment is a much larger topic in engineering ethics in a country such as the United States than in a country such as Germany.

The case of Roger Boisjoly in Chapter 4 shows that even concerns arising from engineers' most fundamental responsibilities may fall on deaf ears. In this chapter we examine various organizations – corporations, government agencies, universities, and research facilities – to see what makes them more able to listen. In Chapters 2 and 3 we considered many problems of engineers in private practice; however, most engineers work as employees and are immersed in organizational cultures that significantly influence their moral lives. Especially important are the practices an organization has developed to respond to "bad news" and to promote fair treatment of employees. Milton Friedman's view that the sole responsibility of managers is to "make as much money as possible" (consistent with obeying the law and conforming to ethical custom[3]) is not generally accepted in management circles today. Current thinking is highly critical of "short-term management," that is, management directed at short-term profits rather than goals of consumer trust and public good will, quality products and service, and high morale and a continued professional growth among employees, factors that generally contribute to *long-term* profitability. Companies devoted to long-term goals are often explicitly concerned to meet ethical standards even in advance of custom. One widely held view of management in a rapidly changing world holds that organizations must continually learn if they are to succeed. This view emphasizes the importance of not only hearing the bad news and anticipating problems before they arise, but also fostering both *personal and* professional growth of employees and responding to employees' desire to build something important, as well as pursuing their own self-interest.[4] Of course, theories are not always reflected in practice. It is crucial that an organization's actual practice not reward managers for maximizing short-term profits or for failing to report expensive problems with consequences that will not be attributed to them.

ENGINEERS AND MANAGERS

A recent study of communications between engineers and managers by researchers at the Center for the Study of Ethics in the Professions at the Illinois Institute of Technology[5] reveals how managers respond to unwelcome news from engineers in well-run high-tech companies. The study identified three value orientations of companies depending on whether the company gave first priority to

- customer satisfaction
- the quality of their work/products
- the financial bottom line

Although this is a rough typology and the priority given those factors are matters of *degree,* for simplicity the report speaks of three types of companies. I shall call the first "customer-oriented" companies, the second "quality-oriented" companies, and the third "finance-oriented" companies.[6] (The identification of both customer-oriented and quality-oriented companies comes as welcome news to some young engineers who fear that concern with the bottom line always dominates other concerns as Friedman argued it should.)

The types of companies differ in several ways: In the quality-oriented companies, quality (and, of course, safety) take priority over cost and the customer's desires. Cost is still considered, but as one engineer put it, "Cost comes in only after our quality standards are met."[7] Quality-oriented companies listen to their customers but take pride in being willing to say "no" to them. In one manager's words, "If a customer wants us to take a chance, we won't go along." Such companies try to convince customers to keep their applications of a product within the company's specifications for its appropriate use, but if they fail to convince the customer, they forfeit the business rather than supply a part or a device that will not perform the customer's job well. Although this strategy does not maximize short-term profits, the quality-oriented companies in this study had secured a large and growing share of the markets in which they competed, so their reputation for quality seems to have contributed to their long-term success.

Even in the quality-oriented companies, managers and engineers had different concerns and priorities. The engineers were likely to see managers as more concerned about cost or more superficial in their judgment and the managers see the engineers as likely "to go into too much detail."

In the customer-oriented companies, customer satisfaction was the main objective. They replaced the internal standard of the quality-oriented companies with an external standard of satisfying the customer. Predictably, in such companies, engineers' quality concerns often conflicted with managers' desire to please the customer.

Davis and his colleagues found both engineers and managers to be critical of finance-oriented companies (perhaps because all of the informants in the study had moved on to customer- or quality-oriented companies.) In finance-oriented companies the desire to maximize the number of units shipped conflicted not only with the engineer's concern for quality, but in some cases even with ethical standards, when engineers or managers were pressured to adjust test results to make it seem that the product met the customer's specifications.

The ten companies in the study [8] were very diverse in respect to size and industry but had some important common characteristics in addition to being all quality-oriented or customer-oriented rather than finance-oriented. With one exception, companies were ones in which communications between managers and engineers were good. (This is not surprising from the way in which the companies were selected for the study.) The study found a high degree of trust that disagreements between engineers and managers over questions of quality and safety could be satisfactorily resolved by simply bringing more people into the decision. The confidence that a genuine consensus can generally be achieved by such means shows a widespread trust in coworkers' competence and the existence of shared goals. In the one exceptional company the engineers were generally undervalued and management presumed to "do too much of the engineering."

An earlier study reported in Robert Jackall's *Moral Mazes*[9] gave a very negative picture of communications between engineers and managers, but this study was done with very different companies and higher levels of management.

An important result in the study reported by Davis, and that ran contrary to the investigators' expectations, was that managers expected engineers to "go to the mat" for safety or quality concerns. This expectation held at the customer-oriented and quality-oriented companies that made up the total of the companies *directly* studied. (Because the companies where financial outcome was the dominant concern were studied only through the reports of engineers and managers who no longer worked for such companies, the evidence about the interest of managers in finance-oriented companies in having engineers fight for quality or safety remained largely unexplored.) Even managers at customer-oriented companies, in which managers expected to sometimes overrule the engineers on matters of quality, if not on safety, wanted to hear the strongest case for quality from their engineers. The engineers studied generally felt their safety judgments were accepted. The managers studied stressed the importance of appreciating the engineers' evaluations to do their own jobs well.

Many factors influence the relationships between engineers or scientists and their supervisors. The relationship of an engineer or a scientist to a supervising manager in a corporation is quite different from the relationship of a research supervisor to a graduate student in engineering or science, for example. Graduate students are somewhat like medical students in their relationship to their research supervisors, except that the relation to the thesis supervisor dwarfs all other supervisee relationships that graduate students have. Both graduate and professional students are in a vulnerable position in case of conflict with those supervisors. Nonetheless, issues not only of quality, product safety, productivity, and customer or sponsor satisfaction but also of laboratory safety, harassment, prejudice, and a hostile work environment are common to university, agency, and corporate settings. University or departmental cultures, corporate cultures, and agency cultures all vary in the support they give or fail to give for raising ethical concerns and in their willingness to monitor or control the activities of their members. Organizations vary significantly in their policies regarding such matters as whether they reserve the right to read employee or student e-mail or the computer files stored on the university, corporate, or agency computers, or whether they subject their members to drug and other biological testing.

The study by Davis and his colleagues of communications between engineers and managers reveals how managers respond to unwelcome news from engineers in high-tech companies. Knowing that even good patterns of communication of the sort revealed in the study by Davis and his colleagues sometimes fail, and recognizing the importance of heeding the warnings of engineers, an increasing number of large U.S. companies have instituted complaint procedures or hot lines to ensure that difficulties are recognized and appropriately addressed and that those who raise concerns in good faith are protected from retaliation.

ORGANIZATIONAL COMPLAINT PROCEDURES

"Complaint procedures" may sound vaguely repellent, since no one wants to be known as a complainer, but it is the standard term for the procedures by which an organization ensures its ability to hear that something is wrong. "Complainant" rather than "complainer" is the term for someone who uses such procedures. If, as Natalie Dandekar has observed, loyalty to an organization ought to mean "loyalty to ethical standards characteristic of the organization at its finest," complainants acting in good faith should be recognized as the most loyal of employees.[10]

Frequently, the occasion for a complaint involves a difference in judgment rather than an accusation of malfeasance. Some disagreements stem from reasonable differences of opinion, some from innocent mistakes; others are due to negligence, or, more rarely, from evil intent. What is morally blameworthy is sometimes not the original mistake but the failure to heed arguments and evidence brought forward to show that a judgment is mistaken. In this way an unresponsive organization transforms innocent mistakes into negligence.

Through complaint procedures, an organization can ensure that bad news is not repressed. Not all complaints are ethically significant, or even well founded. The ethics officer of one large company contractor with a reputation for maintaining high moral standards reports that the majority of complaints that come to his office are about food in the cafeteria. It is not that the food at that company is so bad, but that food is something that people readily complain about. Scattered among the food complaints are matters that really require the attention of the ethics office.

In 1986, in response to public outcry about the high cost of items obtained under defense contracts and about outright financial fraud, defense contractors formed the Defense Contractors' Initiative, which established standards for government contractors to handle employee complaints, including those of financial fraud. As part of this initiative, participating contractors established complaint procedures and an office for handling concerns. This is often called the "ethics office," although it is sometimes called a "compliance office" – the difference in name reflects a significant difference in thinking about its function. Is the office intended to foster ethical conduct or is it meant simply to ensure that the organization is in compliance with regulations?

Safe and effective complaint procedures come in many forms, some formally instituted and others arising de facto. Large organizations may provide separate routes for raising concerns about product safety, laboratory or worker safety, coworker's substance abuse, misuse of funds or fraud, and questions of fairness in promotion or work assignment. In small companies or start-ups the procedures may be entirely informal. Large companies may announce an "open-door policy," in which employees may bypass lower layers of management and take concerns directly to the top, or may employ an "ombuds" or "ombudsman," whose job it is to remain neutral in controversies and to inform complainants of their options or facilitate their exercise. A start-up company in which everyone routinely deals directly with everyone else and good advice is available from many sources may have no need to announce an open-door policy or designate an ombuds. Some companies legitimize the delivery of bad news through "screwup boxes." These work somewhat like suggestion boxes and people may use them anonymously. Complaints are posted on bulletin boards, along with management's responses. Anonymity is a two-edged sword, however, since if the complainant is not identified, the complainant cannot be protected against reprisal from those who guess the complainant's identity. For this reason, many larger companies provide both anonymous and identified means of raising concerns.

Whatever their form, complaint procedures must have certain characteristics if they are to work. Privacy theorist Alan Westin lists characteristics of complaint procedures that make them effective and eliminate the need for whistle-blowing and litigation.[11] Though originally meant for employees within companies, they apply as well to universities, government agencies, hospitals, or other organizations and their employees, students, and trainees.

1. The complaint and appeals mechanism must fit the organizational culture.
2. The means of dispute resolution must inspire general confidence.

3. Top management must display continuing commitment and involvement in the process.
4. The organization must reward merit.
5. Formal procedures must guarantee the process, without creating a legalistic atmosphere.
6. The organization must continually emphasize the availability of channels.
7. Employees must have assistance to bring forward their complaints.
8. Someone must be the advocate of fairness itself, rather than of any particular group or position.
9. All who raise issues or give evidence must be protected from reprisal.
10. Line managers (managers who make decisions central to the work of the company) must support the procedures.
11. The organization must accept the responsibility to change in response to what the process reveals.
12. The organization must, without violating privacy, make public the general nature of the problem, the procedure used to examine it, and the outcome.
13. Probing employee surveys that actively seek concerns must supplement the concerns individuals bring forward.
14. Employee representation must be part of the process.
15. A fair dispute procedure must be established as permanent.

The fit between the complaint and appeals mechanism and the organizational culture is important for the mechanism to be accepted and trusted. For example, the high value placed on academic freedom in a university environment contrasts with the high value placed on obedience in the military. Complaint procedures in a university and in a military organization would necessarily reflect differences in practices of the two types of organization. Even organizations of the same general type may have their own traditions and ways of working that must be taken into account in setting up procedures.

The characteristics Westin identifies as desirable may conflict with one another in a given circumstance. For example, in the mid-1980s many research universities realized that their procedures to deal with complaints of wrongdoing or misconduct in research were inadequate. Complaints of misconduct had been mishandled and retaliation against complainants had occurred. Universities revised their procedures. Those in which decision-making centers in departments often choose to handle at least the initial "inquiry" stage at the department level. Handling the inquiry at the departmental level fits with existing culture and practice, in accord with the first of Westin's characteristics, but is at variance with his fifth and fifteenth requirements. Because charges of misconduct are infrequent, those within a department asked to participate will almost all be doing so for the first time. Even very intelligent and conscientious people doing things for the first time make mistakes. What they learn from one experience does not become institutional learning if each department handles things in its own way and has little opportunity to benefit from the experience of others. Handling the inquiry at the departmental level is at odds with the requirements that there be formal guarantees of the process and that a fair process be established as permanent. That Westin's criteria are useful for evaluating complaint procedures for the experience of many disparate organizations confirms their importance. They do provide valuable design criteria but not a recipe for creating a procedure.

Large organizations usually have established complaint procedures. Before accepting a job one would do well to inquire about those procedures. What does one learn about how they measure up against Westin's criteria? The company's culture? The level of awareness of the procedure by members of the organization? Are employees willing to talk about the company's procedures?: If they say the "open door comes around and hits you in the rear," that is important information, but so is their reluctance to give you their opinion.

Some ethically objectionable situations lend themselves to organizational solutions, others to technical solutions, some to legal or legislative solutions. Some objectionable situations are so egregious that everyone should speak out against them. Others are less grave and best left to those who are familiar with the situation or prepared to address it. In an imperfect world, many things call for reform. Which of those a particular person should work to change, like the question of what profession to choose, or whom, if anyone, to marry, is a question larger than one of professional responsibility. It is a question of what is important in life and what is a person's particular vocation.

GOVERNMENT AGENCIES

Agency culture, good or bad, is often slower to change than that industry, but agencies, like companies, are coming to recognize the importance of employee concerns programs. The Department of Energy (DoE) is among them. It has established an internal avenue to receive all types of employee concerns, to ensure that these concerns are reviewed, referred, or investigated, and to guarantee the person who originates the concern an appropriate response. The DoE program encourages employees to resolve disputes with their first-line supervisor unless that individual is a factor in the concern.

Mixed Results: The Nuclear Regulatory Commission (NRC)

In 1987 the Nuclear Regulatory Commission (NRC) revised its previously existing procedures for dealing with employee concerns. (The NRC is the federal agency charged with oversight for the U.S.'s 110 commercial nuclear reactors, and so is the agency to whom employees of power plants may report safety problems that go unheeded in their home facilities.) The NRC procedures are called the "Differing Professional Views (DPVs) and Differing Professional Opinions (DPOs)" programs. A leading engineer in engineering ethics, Stephen Unger, considered the effects of these changes in the second edition of his book, *Controlling Technology*. Most workers who have used the new procedures would not do so again. Such evidence leads Unger to conclude that the NRC procedures still work poorly.[12] Several of his criticisms focus on the aspects of the procedure Westin finds crucial. Unger finds no general confidence in the means of dispute resolution, and, specifically, no conviction that those who raise issues or give evidence will be protected from reprisal. Westin's twelfth requirement of making public the nature of the problem, the examination of it, and the outcome may be compromised as well, for in the name of national security, the NRC is required to make public only portions of a case. Unger fears that this loophole allows the NRC to cloak blunders or manipulate public opinion.

Given the flaws in the NRC guidelines, it may not be surprising that the NRC has a mixed record of supporting those who report safety violations at nuclear power plants. The NRC was recently found to be ignoring serious safety violations so that nuclear power plants could avoid shutting down or operate more cheaply. George Betancourt and George Gatalatis,[13] two senior engineers at Northeast Utilities, had tried to bring to light dangerous violations at their company's plant Millstone Unit 1 in Waterford, Connecticut. The resistance they encountered from supervisors at Northeast Utilities led them after eighteen months to take the violations to the NRC, only to find that the NRC had been winking at such violations for years. Other engineers at Northeast who had objected to safety violations at the plant had also found little support from the NRC.

American Forestry Service

The handling of differences in professional judgment has been a sore point within the Forest Service as well as the NRC. They both lack a culture that leads "line officers" (the counterpart of line managers in industry) to listen to their technical experts and to develop a complaint procedure with the characteristics Westin describes. As Doug Heiken of the American Forestry Service Employees for Environmental Ethics (AF-SEEE) points out, communications and decisionmaking are especially jeopardized at the Forestry Service because negative consequences are often far removed from the decisions that caused them. This circumstance combined with a policy of holding district rangers accountable only for the effects on a locale during their service there undermines the incentive to make careful decisions. Below is an edited version of a scenario that Heiken gives to illustrate the problem.

THE WRONG INCENTIVES

Suppose that a Forest Service hydrologist finds that the ranger's predecessor boosted timber targets by violating forest plan standards designed for the protection of watersheds and now many of the watersheds on the district are in poor condition. The watershed is healing but could degenerate rapidly if there is greater than normal precipitation in the coming years. If bringing this bad news simply puts the hydrologist and each person who must transmit the information in an unwelcome role, neither the hydrologist nor anyone else will want to pass it on. The hydrologist will not even want to recognize the danger herself. She will have strong incentive to "let sleeping dogs lie" and simply hope the rains won't be too heavy.[14]

The failure to heed evidence about risks does not lead to something like an explosion the following day, as in the *Challenger* disaster. Repressing bad news in the Forest Service does not even lead to an event, like the collapse of a skyscraper, which, even if delayed, will be readily traceable to the technical experts who made the errors that produced the catastrophe. As Heiken points out, judgments such as those to allow cattle to overgraze may have no dramatic effects for years and then a large rainstorm causes noticeable erosion and downstream flooding.

Overcutting of old growth may be widespread on the forest before it is realized that certain wildlife species are rapidly declining. Heiken proposes assigning responsibility to managers for their decisions when the consequences do not become apparent until later, even long afterwards when they have taken a new different position. His general point is that accountability encourages people to deliver the bad news as soon as they learn of it.

These suggestions are very much in accord with Westin's criteria and Unger's observations about the weaknesses in NRC communications. However, as was shown by the engineer–manager communications the IIT study found to work well, certain changes in routine procedures would lessen the need to use complaint procedures, much less to "blow the whistle." These observations are consistent with the finding that in more authoritarian and less responsive organizations, employees have little confidence that internal means of redress will prove effective and so are more likely to become whistle-blowers.[15]

THE HANFORD JOINT COUNCIL FOR RESOLVING EMPLOYEE CONCERNS

One famous whistleblower, Inez Austin, was an engineer employed by Westinghouse at the Hanford Nuclear Reservation, a nuclear weapons facility in Richland, Washington. (As we shall see in Chapter 8, the Hanford weapons facility is the site of the worst radioactive contamination in the United States.) In the summer of 1990 she refused to approve a plan wherein pumping radioactive waste from one underground tank to another would have risked explosion. She was subsequently harassed, sent for psychiatric evaluation, and had her home broken into. Her case brought attention to the abuse of complainants as well as to safety, environmental, and security lapses at the Hanford Reservation.[www] In February 1992, Inez Austin, like Roger Boisjoly before her, was awarded the AAAS Award for Scientific Freedom and Responsibility for her exemplary efforts to protect the public health and safety. After many cases of abuse of complainants who reported threats of a nuclear accident and pollution of the environment with toxic chemicals and nuclear wastes at Westinghouse, Hanford, strong measures were needed to begin to rebuild the trust of employees and of the public.

The Hanford Joint Council for Resolving Employee Concerns was formed after a 1992 landmark study commissioned by Westinghouse Hanford Company and carried out by the University of Washington's

Institute for Public Policy and Management confirmed that severe re-taliation had often followed the raising of a concern at the Hanford facility. Among the study's findings were that every complainant they interviewed was sincere and credible and that Westinghouse's practice of responding to whistle-blowing incidents by commissioning security department investigations of the cases and sending whistle-blowers for psychiatric evaluations was unwarranted. This retaliation included mul-tiple instances of illegal surveillance. In 1991 the inspector general of the Energy Department, John C. Layton, had found an array of sophisti-cated eavesdropping equipment in the possession of Westinghouse and Battele, another contractor at the Hanford site, and at other weapons sites in Idaho and South Carolina. Only state or congressional law en-forcement agencies are allowed to use such surveillance equipment.[16]

The Hanford Joint Council[www] began considering cases in January 1995. It is made up of three public interest representatives: two unaf-filiated neutral parties; an ex-whistle-blower; and two managers from Westinghouse, the contractor for the Hanford Nuclear Reservation.[17] The council is a chartered, nonprofit organization with no legal ties to Westinghouse Hanford or the Department of Energy. However, it has full endorsement from Westinghouse Hanford Company, the State of Washington Department of Ecology, and the U.S. DoE[18] and a commit-ment from the DoE to operate for a minimum of five years with the flexibility to expand its mandate and hear concerns from employees of other contractors from the Hanford site.

The Hanford Joint Council is an innovative attempt to restore public trust and secure effective cooperation in accomplishing a difficult and dangerous cleanup. Walter Elden, who helped create the IEEE Member Conduct Committee, on which he now serves, cites the Hanford Joint Council as a model for supporting an engineer's fulfillment of their re-sponsibilities for public health, worker safety, and environmental preser-vation in the face of pressure of financial concerns, time constraints, or political expediency.[19] It is a model worth emulating, although it would be a mistake to believe that the Hanford Joint Council can immediately curtail harassment in a working environment that has degenerated as far as the Westinghouse Hanford facility had.[20] We have seen the intimate connection between professional responsibility and trustworthiness: To carry out the ethical responsibilities that go with being an engineer, en-gineers need not only be trustworthy, but must have trustworthy means of bringing forward concerns that arise out of their engineering knowl-edge and experience. Restoring trust after it has been seriously betrayed is a difficult matter. A review body such as the Hanford Joint Council, which represents all factions, provides one important mechanism for restoring trust.

EMPLOYMENT GUIDELINES FROM ENGINEERING AND SCIENTIFIC SOCIETIES

For a new engineer or scientist the first chance to gain an impression of the organizational culture of a potential employer, other than a summer job, is usually the job interview. The "Guidelines to Professional Employment for Engineers and Scientists,"[www] available from the IEEE and other signatory organizations,[21] provides a benchmark or normative set of expectations on both employees and employers. The guidelines are useful for assessing the ethical climate at a potential employer and for understanding one's own obligations in conducting a job search. These guidelines have received wide endorsement by about thirty professional societies of engineers and scientists beginning with the NSPE, the American Institute of Chemical Engineers, American Society of Civil Engineers, American Society of Mechanical Engineers, and the Institute of Electrical and Electronics Engineers. Many employers as well as employees contributed to formulating these guidelines. They make good reading in preparation for transition from college to the work world and make an appropriate topic for discussion in an extended job interview.

The objective of the guidelines is to help professional employees and employers to establish a climate that enables them to fulfill their responsibilities and obligations. The guidelines list a wide range of responsibilities of professional employees and their employers. These responsibilities are grouped under four headings: recruitment, employment, professional development, and termination and transfer. The topics they cover range from those already familiar from engineering codes of ethics to expectations about performance reviews.

The guidelines identify the following as prerequisites for an ethical climate that supports the fulfillment of responsibilities:

1. A sound relationship between the professional employee and the employer, based on mutual loyalty, cooperation, fair treatment, ethical practices, and respect.
2. Recognition of the responsibility to safeguard the public health, safety, and welfare.

3. Employee loyalty and creativity in support of the employer's objectives.
4. Opportunity for professional growth of the employee, based on employee's initiation and the employer's support.
5. Recognition that discrimination due to age, race, religion, political affiliation, or sex should not enter into the professional employee–employer relationship. There should be joint acceptance of the concepts which are reflected in the Equal Employment Opportunity regulations.
6. Recognition that local conditions may result in honest differences in interpretation of and deviations from the details of these Guidelines. Such differences should be resolved by discussions leading to an understanding which meets the spirit of the Guidelines.[22]

The guidelines are intended to draw clear lines where clear lines can be drawn and to give general guidance in the many areas where discretion must be exercised. They define abuses, such as the acceptance of one job and then refusing the job to take a later offer or summarily rescinding a job offer that an applicant has accepted. (On this point the guidelines state, "Having accepted an offer of employment, the applicant is morally obligated to honor the commitment unless formally released after giving adequate notice of intent," and "Having accepted an applicant, an employer who finds it necessary to rescind offer of employment should make adequate reparation for any injury suffered.") They also disapprove of such practices as employers forcing employees to promise that if they leave their company they will not go to work for a named competitor. This disapproval is balanced with a strong statement of an employee's obligations not to divulge proprietary knowledge of a former employer. (The guidelines say "Agreements among employers or between employer and professional employee which limit the opportunity of professional employees to seek other employment or establish independent enterprise are contrary to the spirit of these Guidelines" and "The applicant should carefully evaluate past, present, and future confidentiality obligations in regard to trade secrets and proprietary information connected with the potential employment. The applicant should not seek or accept employment on the basis of using or divulging any trade secrets or proprietary information.") Although it is clear that one should not tell trade secrets of one employer to another, there are many subtle issues about proprietary information where drawing the line is not so easy. How much know-how acquired in one job can one take to another? If that know-how is proprietary does that mean one should do work in a way that one knows to be inferior? Should one refuse to work in an area similar to that which one had worked at a previous employer? Should one refuse to work for a competing company? These are complex matters that cannot be answered with simple rules.

ORGANIZATIONAL CONTROL AND INDIVIDUAL PRIVACY: THE BIOLOGICAL TESTING OF WORKERS

Some matters of the employer–employee relationship remain quite controversial and are not covered in the employment guidelines just discussed. One of the most controversial areas is that of sacrificing the personal privacy of employees to an employer's need or desire to have information about the employee. As was remarked earlier, organizations vary greatly in policies on such matters as whether they may read employee or student e-mail or the computer files that students or employees store on the university, corporate, or agency computers. [23] Biological testing is another area of conflict between an organization's interest in knowing and an employee's right to privacy. [24] On the horizon is the question of whether companies should be allowed to use DNA and other biological testing to exclude from the workplace workers who would be genetically predisposed to occupational diseases associated with contaminants in their workplace.

The refusal in 1995 by two young U.S. marines, John Mayfield and Joe Vlacovsky, to give DNA samples for a military DNA registry brought national attention to DNA data banks. (The two were court-martialed for disobeying a direct order. The final disposition of the case was that the two marines were given a reprimand and confined to the base for seven days, but they received honorable discharges, full veterans benefits, and kept their DNA. [25]) The U.S. military started this data bank in 1991 and planned for it to contain DNA samples from all enlisted personnel.

Because the infringement of privacy rights with which engineering students are most likely to be familiar is drug testing by employers or potential employers, that is the example we will consider in some detail. Drug testing raises issues of both the justification for acquiring information from the test and the demeaning circumstances of the testing procedure. The questions are separable even if a reliable test result requires testing conditions that are somewhat demeaning. (The test is clearly not worth doing if results can be easily faked, but the measures taken to ensure that results are not faked, from shutting off the water in the restroom sinks, to the more stringent measure of having the giving of the sample witnessed, make the test more demeaning.) Why *do* organizations from the armed forces to high-tech corporations want to conduct drug testing, and when, if ever, is the infringement of privacy justified? If justified, how is the test best conducted to be fair and considerate to those affected?

The first and most obvious justification given for screening for drugs (that is, drug testing of persons an organization has no reason to think are using drugs) is that subtle impairment of tested workers endangers the public safety. The argument from public safety is plausible only for occupations in which response time and coordination, affected by small doses of psychoactive substances, are critical to the safety of others. Air traffic control, truck driving, surgery, and piloting aircraft are such occupations. Although engineers and scientists sometimes perform work in which response time and coordination are critical, that is not the norm. Let us look for possible justifications of the general testing of the engineering and science work force.

Apart from the hazard to public safety from substance abuse by workers in safety critical jobs, substance abuse by any workers significantly reduces productivity. The Research Triangle Institute in North Carolina estimates that in 1983, productivity losses due to drug abuse were $33 billion and losses due to alcoholism were $65 billion. Health care costs resulting from this substance abuse are estimated at another $9 billion. (These figures do not include the even greater costs to the substance abusers and their families.) The loss to companies takes many forms, from absenteeism, which is four to eight times higher for alcohol abusers than for nonabusers, to fatal accidents, 40% of which are attributable to alcohol abuse.[26] Furthermore, companies can be held legally liable for what an affected employee does outside the workplace. In one case an employer was found liable for sending home an intoxicated employee who while driving home hit two people and killed them.[27]

Clearly, employers have an interest in reducing productivity losses. Drug testing helps to identify substance abusers. Furthermore, testing deters drug use. Drug testing can be done on urine, blood, saliva, breath, hair, or brain waves, but urine tests are most common.

Tests are imperfect and each has **false-negative** and **false-positive** rates associated with it. These are rates at which the test, even though correctly administered, will give the wrong indication. The false negatives are instances in which traces of a drug or its characteristic metabolites are in the sample tested but not detected in the test.[28] False positives are instances in which the test result is positive but due to factors other than drug use. An important issue of fairness, in addition to that of the fairness of testing without cause, is that of appropriate protection for employees who might have false-positive results. In one notorious case, two members of the armed forces were court-martialed and dishonorably discharged after testing positive for opiates. It was eventually established that their positive test was due to the poppy seeds on bagels they had eaten the morning of the test.

To some extent, false positives can be traded off against false negatives in a test by adjusting the concentration of a detected substance that counts as a positive result. As one might expect, the tests that have low false-positive and false-negative rates (that is, are best in identifying those and only those who are using drugs) are also the ones that are more costly to use. Most companies that conduct random or universal testing of their employees for drug use test all samples with an inexpensive test, such as the EMIT (Enzyme Multiple Immunoassay Test), and then retest any positive samples with a more expensive test, such as gas chromatography. This reduces the likelihood of a false positive without incurring the risk of a false negative.

For established employees, the usual consequence of a first-time positive drug test is referral to an "employee assistance program." Employee assistance programs are intended to help with the many problems and traumas that can interfere with job performance. Substance abuse, personal problems, marital difficulties, worry about one's own or family member's health, and bereavement at the death of a friend or family member are among the situations that can lead employees to go to their company's employee assistance program. Many are problems that people cannot entirely avoid. Employee assistance programs offer counseling and other services to help workers cope with such difficulties and suffering. Ought an employer screen for these difficulties too, or is there a justification for screening only for substance abuse?

For good or ill, few companies take steps to monitor their work force for problems other than substance abuse. Most large companies (and all government contractors) provide employee assistance programs, however. Referring people to employee assistance usually occurs after their work performance falters, although it might occur simply because a person is visibly distressed. Complete confidentiality is required for employees to trust using such programs. If the content of any session were not protected by confidentiality, employees would have the experience of being under surveillance, which would only add to their stress and

WORKPLACE RIGHTS AND RESPONSIBILITIES

make them reluctant to use the service. For this reason, many companies subcontract employee assistance services so that the counselors in the program are not company employees. When employees are referred to an employee assistance program, it is usually up to the employees to follow through, although in the case of referral for positive drug tests, employees might be required to demonstrate that they did go to the service.

Except for the case of drug testing, few employers screen their employees for signs that they are in difficulty. Indeed, some companies subject employees to drug testing only if they gave reason to believe they have been abusing drugs, but other companies subject their employees to random or universal drug screening, and many subject job candidates to a preemployment drug test. What, if any, justification is there for putting drug use in a special category and screening employees for it?

> "Screening" is testing of a large number of individuals designed to identify those with a particular characteristic or biological condition. Random or universal testing constitutes screening. Testing a person because there is reason to believe he has some condition does not.

If it proposed to do random or universal drug testing of the citizenry, the U.S. government would be in violation of the Fourth Amendment to the Constitution (in the "Bill of Rights"), which states:

The right of the people to be secure in their persons, houses, papers, and effects, against unreasonable searches and seizures, shall not be violated, and no warrants shall issue, but upon probable cause, supported by oath or affirmation, and particularly describing the place to be searched, and the persons or things to be seized.

The requirement of probable cause for a warrant to search someone's person is met if that individual gives signs of being under the influence of drugs or gives evidence of having used drugs on the job. Arguably at least, it is met by the policy, used by some companies, of testing those involved in accidents. It is not met in random or universal screening of employees.

Drug testing as a condition of employment is different from government screening of citizens in that potential employees have some freedom not to seek employment with certain companies. (Recall that one offensive feature of Ford's treatment of Todd Riggs was its failure to inform him that drug testing was a part of the plant trip. If the company had informed him, he could have decided if he wanted to take the trip under those conditions.)

Companies commonly require a preemployment physical exam. Is the requirement of a preemployment physical and a preemployment drug test to be viewed similarly? Is either justified?

To answer these questions, we must know about the substances tested for, the accuracy of the tests, and the conditions of testing. Illegal drugs are almost always included, and any of a variety of prescription medications may also be tested for. (Some substances are common to illegal drugs and prescription medications.) Some people argue that managers may have a right to know about illegal drug use, but not to monitor the details of one's health care, and hence they find the testing for prescription drugs more objectionable. The Americans for Disabilities Act protects workers with disabilities. Some workers require medications to cope with their disabilities. Does testing for such medications put them at special disadvantage or invade their privacy in a way that is discriminatory?

Some question a company's right to test for *illegal* drugs arguing that it is intrusive for an employer to go looking for legal violations that occur outside work time, when most detectable drug use occurs. The length of time that substances remain detectable in a user's urine varies greatly. Alcohol lasts not much longer than the significant influence on performance (from four to twelve hours), cocaine is detectable for two to four days, and marijuana for several weeks (depending on use). Drugs like barbiturates, valium, and darvon are somewhere between the extremes of alcohol and marijuana. Drug testing might pick up marijuana used weeks earlier, perhaps while on vacation, and miss the alcohol abuse that affected on-the-job performance earlier the same day.

Would it be acceptable for an employer to go looking for other illegal activities by employees, say by searching records of unpaid parking tickets or using surveillance to discover employees' illegal gambling? Are such examples closely analogous to screening for illegal drugs?

One justification given for drug testing is that parties who receive government grants and contracts, as do many employers of engineers, must provide a "drug-free" workplace. Universities also receive such grants and contracts and make such certifications, however, without random or universal testing of their employees for drugs.

As we have seen, alcohol causes the greatest losses to employers, yet it is one of the substances that is detectable for the briefest time. Alcohol intoxication is also more readily observable than marijuana intoxication. Why not alert managers to the signs of alcohol and substance abuse, and when they find probable cause, *then* require testing? Substance abuse problems, like anxiety and depression and other "employee assistance" problems, prove difficult for many managers to raise. Rather than do intensive management training on this subject, many companies simply take a technological fix and test their employees randomly or universally. It will be interesting to see whether the requirement that supervisors (as well as managers) deal appropriately with issues of sexual and other harassment will increase the competence of supervisors to handle touchy issues and open the way for personal rather than technological responses to substance abuse.

Some companies have respected employees' or recruits' objections that drug screening is an invasion of their privacy. In some cases, companies have waived requirements for preemployment drug tests for students and recent graduates who have argued that without probable cause such a body search is demeaning and unjustified. When one high-tech company instituted random drug testing some years ago, a respected employee announced that he found such testing an affront and would never submit to drug testing. Although refusal to take the test is supposedly grounds for dismissal at this company, in the eight years that this drug testing has been in place the employee's name has never come up despite the supposed randomness of name selection.

> Some landmark legal decisions in California said that employees had rights of recovery against their employer if fired precipitously if they refused to take a drug test. Thus the right of an employer to test is hedged with many countervailing employee rights.

Some of the arguments offered for testing have not had ethical justifications but have been only pleas that what might be ethically desirable is not practical, at least as things now stand. Such arguments raise the further question of whether the present situation (including organizational practices, the technical limits of existing tests, and legal guarantees for employers and employees) should, as a moral matter, be reformed, and if so, in what way.

LIMITS ON ACCEPTABLE BEHAVIOR IN THE LARGE CORPORATION

Companies, even those that screen their employees for drugs, do not simply put their interest in increasing productivity ahead of such employee rights as privacy. In fact, ethics training at reputable companies encourages employees to consider their own rights and interests, so long as furthering them does not exclude other important considerations. The responses to problems that such companies favor are not the most self-sacrificing, but rather those consistent with the following criteria:

1. getting to the root of the problem
2. protecting the public's interest, especially in health and safety
3. showing respect for the law
4. keeping the company honest
5. protecting the company's reputation for honesty and fairness
6. promoting good working relationships among people in the company
7. making appropriate use of organizational channels
8. reducing aggravation to that required to meet the other criteria

These are practical considerations that serve to help the company flourish. Considerations such as protecting the company's reputation and making appropriate use of organizational channels have ethical implications: Fulfilling them helps to maintain the trust both inside and outside the company that is necessary to meet other, more obviously ethical, criteria. Ethics training in reputable high-tech companies often emphasizes developing the knowledge and discretion to design responses to meet all these criteria if possible, so "everyone wins."

The ethics materials from two large high-tech companies, both government contractors, Lockheed Martin and Texas Instruments,[www] are good examples of such training. *Gray Matters*, for instance, is a game that companies use to teach ethics. It was written by George Sammet for Martin Marietta, now Lockheed Martin, and has since been used by other high-tech companies, such as Boeing, Honeywell, McDonnell Douglas, and General Electric.

Lockheed Martin's Ethics Game

The game consists of over a hundred mini-cases that very briefly present ethically significant situations that call for a response. These range from observing a coworker snorting cocaine, to being instructed to mischarge your time, to communicating with subcontractors. The point of the game is to make employees aware of ethical problems that can arise in their day-to-day responsibilities and to enable them to think through the consequences of their decisions and actions.

Each mini-case is accompanied by four potential answers. (Usually the game is played in groups. The group discusses the case and comes to a decision about which answer provides the best course of action.) The answers are scored (from −20 to +15), and an explanation or a rationale for the score is provided. The potential answers and evaluations of those answers inform employees about the company's values and standards on business ethics, develop their skills in applying company standards, and help employees find the proper procedures for addressing a variety of ethical concerns within their company.

Consider mini-case 68:

> You have been assigned to work on a proposal to the government. The proposal manager tells you and several other nonexempt workers that he'd like you to stay home Thursday and Friday and then come in and work Saturday and Sunday but to report that you worked on Thursday and Friday. That way, you would work 40 hours for the week, but the company would not have to pay you overtime for the weekend. "After all," he says, "proposal money is short."
> What do you do?

The answers offered are:

A. Grudgingly comply thinking these days a job is a job.
B. Check with Human Resources to see if company policy permits this.
C. Call the ethics officer and allege unfair treatment.
D. Speak up immediately and question the manager's right to impose such a condition.

Answer A is scored −5 points with the comment "To go grudgingly along with a company imposition is not conducive to good morale. Isn't there a better way?" The negative points indicate that this is a mildly bad answer and the comment discourages allowing oneself to be exploited.

Answer D, directly challenging the supervisor, is scored +5 points, meaning that it is a moderately good answer, with the comment "Certainly you are within your rights to do this."

Answer B receives the highest number of points of this set, +10 points. "This is a sensible approach. If company policy doesn't permit this, it will be corrected. If it does, you have the facts needed to make a decision. Most companies and the government would consider this falsifying your timecard, thus denying the manager's right to ask you to do it." Employees learn from this both that the practice requested by the proposal manager counts as falsifying your timecard (and thus is a form of financial fraud) and that some companies may nonetheless allow the practice. Presumably, if your company did permit this practice, you could complain to the government. However, you may not want to be blowing the whistle over as small a matter as this. That your company allows this chiseling does tell you something important about the ethics of your company, however. Response B is the course of action that will get you information most quickly and with a minimal risk to yourself. The remaining answer, C, "Call the ethics officer and allege unfair treatment." is scored +5 points with the comment "This response will take longer, but it will eventually arrive at the same answer as 'B'."

Mini-case 15 is:

A coworker is injured on the job. You are a witness and could testify that the company was at fault. What do you do?

The answers they provide are:

A. Don't get involved.
B. Contact the injured coworker and offer to appear on her behalf.
C. Report to the company what you saw to ensure that the safety hazard is corrected.
D. Protect the company by refusing to appear as a witness for the injured.

Not surprisingly, both answers A and D are scored −10 points. Answer A is termed a cop-out. Unfairness to the coworker is cited in the comment on D. B receives +5 points for showing compassion but doing nothing about the unsafe condition, and C is scored +10 with the comment "Gets at the cause of the injury. Whatever happens after that, happens. If the injured wants you as a witness that is both your rights."

Mini-case 4 addresses the issue of exploitation by coworkers:

For several months now, one of your colleagues has been slacking off, and you are getting stuck doing the work. You think it is unfair. What do you do?

The candidate answers are:

A. Recognize this as an opportunity for you to demonstrate how capable you are.
B. Go to your supervisor and complain about this unfair workload.
C. Discuss the problem with your colleague in an attempt to solve the problem without involving others.
D. Discuss the problem with the human resources department.

Answer A is scored 0 with the comment that although this may solve the workload problem, if you hold up, it does not address the ethical issue (equitable distribution of work) and so receives no positive points. Answer C is presented as the best answer as an initial response, although, as the comments make clear, it might not work, in which case you would have to "take the next step." That next step presumably is B, which is given +5 points, because this strategy brings your colleague's behavior to your supervisor's attention and may give you expanded responsibility. Answer D is scored −5 with the comment "Pushes the problem solving onto someone else. The problem is between you, your supervisor, and your colleague. Solve it there."

The scoring of answers to other mini-cases makes clear the limits on what you can do for yourself and your friends. For example, consider mini-case 14:

> A friend of yours wants to transfer to your division but may not be the best qualified for the job you have open. One other person, whom you do not know, has applied. What do you do?

Scoring and comments discourage putting your friend's wishes or your preference to work with someone you know and like ahead of the company's interest in finding the most qualified person. (This case also teaches employees how to make use of human resources to help with the selection of the appropriate person.)

Finally, on the topic of whistle-blowing, several mini-cases make it clear that just as it makes a great difference who within the company you tell about a particular problem, so it makes a great difference to *whom* you go when taking concerns outside the company.

Consider mini-case 58:

> You are working on a government contract and are convinced that a serious mischarging incident has occurred. You also believe that it was deliberate since the program was running out of funds.

The candidate answers are:

A. Call the Department of Defense hot line.
B. Inform the local newspaper of your suspicions.
C. Discuss it with your local audit office.
D. Send an anonymous note to your corporate ethics office.

Answer A is scored 0: not wrong but the company would rather that you reported it within the company first. The office that is best equipped to deal with the complaint is the local audit office (so answer C gets +10). An anonymous note to the ethics office will get the situation investigated, but anonymity will slow the investigation (+5). The scoring implies that although some wrongdoing may occur within the company, the company commits itself to providing employees with appropriate means of reporting the misconduct without undue risk to themselves. If fraud were a regular occurrence and the company did little to stop it, the situation would be different, ethically speaking; then going straight to the DoD would be appropriate.

Going to the press is scored −10. Telling the press (unlike going to the DoD) elicits a comment about breaching confidentiality. The issue is better described as one of loyalty in giving the company a chance to remedy the situation, rather than confidentiality, since misdeeds do not automatically become confidential matters. Apart from embarrassing the company, going to the press is a less reliable means of getting the abuse addressed than going to the audit office or to the DoD, unless the DoD hot line as well as the company audit office is failing in its function.

Taken as a whole, the mini-cases give a picture of both the ethical values of the company that regularly uses them and of their understanding of the characteristics of a well-functioning work community. The *Gray Matters* game consistently encourages employees to maintain their moral integrity and not allow themselves to be exploited, as well as to be honest and fair with others inside and outside the company. The recommendation is to go to the appropriate offices, such as human resources, the legal department, and the local audit office, for expert advice rather than simply choosing the most self-sacrificing course.

The *Gray Matters* material counsels behavior that is in the company's long-term interest, much as the NSPE code and the judgments of the NSPE Board of Ethical Review show distinctive concern about the collegial relations among engineers. Although *Gray Matters* generally recommends getting information when one is uncertain, it also recommends addressing problems directly when possible rather than referring the problem to someone else. Readers may be surprised at how much information the material instructs employees to take to their supervisors: for example, that one saw a company quality manager snorting cocaine at a Saturday night party (mini-case 23) or that the reason a person was ill and missed work was that he was hung over from partying (mini-case 3). Readers may not agree with *Gray Matters*' consistently negative view of warning one's coworkers about some perceived unfairness on the part of one's supervisor. However, the rules implicit in the *Gray Matters* judgments, including the rule against gossip, find some justification in fostering a work community able to achieve corporate objectives. The cases provide a benchmark set of expectations for an ethically concerned company. Like the guidelines on employment discussed earlier in this chapter, the *Gray Matters* mini-cases might prove useful in discussion with potential employers to see how they would recommend handling the same issues.[29]

Advice from the Texas Instruments Ethics Office

The ethical values and organizational arrangements assumed in the advice from the Texas Instruments (TI) Ethics Office[www] are very similar to the values and organizational arrangements implicit in the answers and scoring of Lockheed Martin's ethics game. In both cases, much of the ethical content concerns responsibility of employees for maintaining a well-functioning work community. Sometimes such concerns are called "etiquette," but like "netiquette," many have ethical significance (unlike the etiquette of which fork to use). Many have an ethical component: If the norms are frequently violated, then major moral responsibilities become difficult to fulfill.

The advice format of the TI Ethics Office articles leaves room for detailed explanation of policies and reflection that finds no place in an ethics game. One of the most interesting of the TI advice articles presents situations (presumably similar to ones that had been reported to the Ethics Office) that looked suspicious but turned out to be innocent. The ethics officers make three points:

1. A situation is not always what it seems.
2. The question of how an action will appear is [often] one that an agent should consider.
3. The appropriate thing to do is report the matter to the ethics office rather than gossip about it or ignore an apparent misdeed. Then the truth can be discovered and wrongdoing stopped or suspicion dispelled. *There is no penalty for reporting something that turns out to be innocent.*

Reporting a matter rather than gossiping about it clearly requires a trustworthy and risk-free means for handling employee complaints and concerns. TI's advice on the scope and limits of benchmarking and reverse engineering in engineering practice is especially important to engineers.

The TI Office offers "ethics quick tests" to assess potential actions. The first four of these are:

1. Is the action legal?
2. Does it comply with our values?

3. If you do it, will you feel bad?
4. How will it look in the newspaper?

Strictly speaking, these are not tests for whether an action is ethically acceptable. In particular, conformity with some companies' values might even lead to unethical behavior, and, as we saw in the introduction, feelings may be more a function of one's personal history than of the ethical acceptability of an act. Such tests are better described as ways to see if an action needs further scrutiny.

THE WORK ENVIRONMENT: ETHICAL AND LEGAL CONSIDERATIONS

Engineering codes of ethics frequently include prohibitions against prejudicial treatment of others. For example, the IEEE code includes a pledge to "treat fairly all persons regardless of such factors as race, religion, gender, disability, age, or national origin." In addition to such ethical requirements, legal constraints and company culture govern the work environment.

Consideration of the law enters a company's assessment of behavior both because of the prudential considerations about avoiding legal penalties and because of the moral authority of law that we discussed in the section on ethics, conscience, and the law in Part 4 the introduction. Norms concerning the work environment are found both in legislation and in case law. Many legal norms are framed in terms of a person's right not to be subjected to an abusive work environment. Legislation itself is often prompted by a change in public consciousness which follows some extreme events. One such event that was particularly significant for engineering schools in Canada and the United States was the "Montreal Massacre." It reawakened awareness to the problem of the high level of violence directed against women and strengthened Canadian public support of gun control.

THE MONTREAL MASSACRE

On December 6, 1989, Marc Lepine shot and killed fourteen women at the Ecole Polytechnique in Montreal using a semiautomatic Sturm Ruger Mini-14 rifle. He also wounded thirteen others, mostly women, before committing suicide with the same gun. All but one of the slain women were students in the engineering school. Lepine blamed his own failures on feminists.

Groups at engineering schools in Canada and the United States hold candlelight vigils on December 6, between 5 and 6 p.m., the time the killings took place, to help ensure that the public does not forget the killings so that such events will not be repeated.

The standards of behavior in ethically concerned organizations, such as universities, corporations, and government agencies, go beyond efforts to comply with the law. Responsible organizations work to promote a positive and mutually respectful work environment. This positive strategy not only helps prevent problems degenerating to the point of being notorious violations of legal rights but also fosters high morale and helps groups function well.

Title VII of the U.S. Civil Rights Act of 1964

The U.S. Civil Rights Act of 1964 grew out of public outrage following the abuse and murders of African-Americans involved in the civil rights movement. Title VII of the Civil Rights Act makes it unlawful for an employer to discriminate against any individual with respect to compensation, terms, conditions, or privileges of employment because of that person's race, color, religion, sex, or national origin. (Title IX makes similar provisions for educational institutions.)

In a 1986 decision (*Meritor Savings Bank v. Vinson*) the Supreme Court interpreted this language to prohibit discrimination that caused injury other than economic. It held that the phrase "terms, conditions, or privileges of employment" shows that Congress intended "to strike at the entire spectrum of disparate treatment" in employment, and that included subjecting people to a discriminatorily hostile or abusive environment. In that opinion the Court had said that mere utterance of an offensive epithet does not so significantly affect the offended employee's working environment as to violate Title VII but that in this case Title VII was clearly violated because the offensive behavior was "so heavily polluted with discrimination as to destroy completely the emotional and psychological stability of minority group workers."

Title VII of the Civil Rights Act and its interpretation in *Meritor* set the stage for the 1993 Supreme Court decision in *Harris v. Forklift*, which further clarified what constitutes a work environment so discriminatorily hostile or abusive as to give grounds for legal action.

U.S. Supreme Court Decision on Harris v. Forklift

Teresa Harris had worked as a manager at Forklift Systems, Inc., an equipment rental company, from April 1985 until October 1987. During the time of Harris's employment at Forklift, Charles Hardy, Forklift's president, often insulted her because of her gender and often made her the target of unwanted sexual innuendoes. For example, Hardy told Harris on several occasions, in front of other employees, "You're a woman, what do you know?" "We need a man as the rental manager." At least once, he told her she was "a dumb ass woman." Also in front of others, he suggested that he and Harris go to a motel to negotiate Harris's raise. Hardy occasionally asked Harris and other female employees to get coins from his front pants pocket or threw objects on the ground in front of them and asked them to pick up the objects.

The Supreme Court's decision in case number 92-1168, Teresa Harris, Petitioner v. Forklift Systems, Inc. (November 9, 1993) further refined the criteria for what offensive behavior constitutes grounds for legal action under Title VII. As it had established in the Meritor case, offensive behavior that causes significant psychological injury is "actionable" (that is, it warrants legal redress). In Harris's case, where the offensive behavior was not as extreme as in Meritor, the psychological injury was not shown to have *disabled* Teresa Harris as it had the complainants in *Meritor*. On this ground the District Court held that Harris's injury was not sufficiently serious to be actionable.

The Supreme Court overturned the lower court decision, stating that it is sufficient that a reasonable person find the work environment hostile or abusive and the victim perceives the environment to be abusive. The Supreme Court held that a discriminatorily abusive work environment can, and often will, undermine an employee's job performance, or keep employees from advancing in their careers, even if it does not cause a nervous breakdown or other severe psychological disability. Discriminatory behavior that creates a work environment abusive to employees because of their race, gender, religion, or national origin violates the norm of workplace equality set out in Title VII. The Court held that whether an environment is hostile or abusive can be determined only by looking at all the circumstances and not solely at the degree of disability that results from it. The factors they mentioned are frequency of the discriminatory conduct; its severity; whether it is physically threatening, humiliating, or merely offensive; and whether it unreasonably interferes with an employee's work performance.

In mid-August 1987, after Harris complained to Hardy about his conduct he claimed he was only joking, apologized, and promised to stop. But in early September, Hardy began anew. On October 1, Harris collected her paycheck and quit. She then sued Forklift, claiming that Hardy's conduct had created an abusive work environment for her because of her gender.

The United States District Court for the Middle District of Tennessee judged that some of Hardy's comments offended Harris, and would offend any reasonable woman, but they held Hardy's conduct would not have risen to the level of interfering with Harris's work performance. In a decision delivered by Justice O'Connor a unanimous court overturned the lower courts' decisions. Justices Scalia and Ginsberg filed concurring opinions.

Extreme cases of abuse more easily draw attention precisely because they are extreme. However, such cases can leave the mistaken impression that only such extreme behavior is objectionable or illegal. *Harris v. Forklift* establishes that evident harassment violates the law even if it does not psychologically cripple the harassed person.

From Overcoming Prejudice to Valuing Diversity

MIT Ombudsperson Mary Rowe has argued that persistent acts of subtle discrimination, most of which is not amenable to legal control, does greater damage than the clearly offensive but rarer behavior of Hardy in the Forklift case.[30] "Micro-inequities," as Rowe calls them, function "like the dripping of water; random drops themselves do little damage; endless drops in one place can have profound effects." These inequities may take the form of persistent application of negative stereotypes to individuals despite their actual attributes. Rowe suggests using measures such as employee attitude surveys to bring out such problems, means for individuals to obtain confidential advice and support, and management training to overcome subtle discrimination. Her suggestions accord with those of Westin, which we enumerated at the beginning of this chapter.

Organizations that want to thrive in a time of an increasingly diverse workforce often run diversity training workshops for their employees as well as mangers. Some go beyond combating prejudice and discrimination to valuing diversity. Valuing diversity sets the goal of recognizing and valuing differences rather than just treating everyone the same. Issues of prejudice are most often framed as issues of "discrimination." However, discrimination, recognizing the differences between things, is often good. It is a compliment to say a person has "discriminating tastes," for example. Unjustified differences in treatment on the basis of race, gender, ethnicity, religion, nationality, sexual orientation, and disability are also called "discrimination" for short. This sort of discrimination leads us to focus on differences in treatment and ask whether those are justified. An equally important question is whether everyone is treated and expected to act the same way, but that that way is one only those in the dominant group finds natural or easy.

A scenario by Joel Palacios (MIT '96) aptly illustrates the difficulty for various minorities when only the majority way of doing things is acknowledged.

DIVERSITY AND BARRIERS TO ADVANCEMENT

Casper and Pepe started working for the same company, at the same time and under the same supervisor, Mr. Harmless. Casper and Mr. Harmless are both European-Americans while Pepe is a Mexican-American.

Soon after they started working, Mr. Harmless invited both of them to his traditional Sunday afternoon barbecue, an event that is held bi-weekly and attended by many of the other professionals in the company. Both Casper and Pepe attended the event. Although Casper seemed to have a great time, Pepe felt uncomfortable because he was the only minority present out of about six employees and their families. His cultural expectations of the event had differed from those of the others. For example, he prepared a dish to share with everyone. Other families had each brought their own food and drinks. He also felt that it was difficult to find common ground with his coworkers outside the world of their profession. Pepe decided not to attend any future barbecues simply because he felt uncomfortable.

The event continued and the supervisor continued to invite both subordinates. Casper attended every time; however, Pepe never did, and he struggled to come up with reasons why he could not do so. He did not want the supervisor and other employees to take his rejections personally. As time went on, Pepe sensed the personal relationship between Casper and Mr. Harmless developing into a strong one. Eventually, a year after they had both joined the company, Casper had received a good promotion, mainly as a result of a fine recommendation from Mr. Harmless.

Pepe had occasionally thought that Mr. Harmless exhibited favoritism toward Casper due to their personal relationship, and he believed this was the main reason for the recommendation. Pepe felt that he had been doing superior work and that his contributions to the company were at least as significant as Casper's. Pepe became even more concerned about the situation when the new subordinate, hired to replace Casper, turned out to be another European-American. A month after this, the new subordinate seemed to be following the footsteps of Casper, developing a strong personal relationship with Mr. Harmless.

What should Pepe do?[31]

ORGANIZATIONAL RESPONSES TO OFFENSIVE BEHAVIOR AND HARASSMENT

Ethically active organizations concern themselves with subtle issues of harassment and work environments as well as with legal issues. These subtler norms are well conveyed in the scoring of two other mini-cases from Lockheed Martin.

Complaints of sexual harassment are most commonly brought by women, although men also make this complaint. Mini-case 57 describes a situation that most often would more typically be experienced by males.

You are a quality inspector. After making your own calculations, you disagree with your supervisor about whether the quality of the item is at an acceptable level. With a rolled up newspaper in his hand, your supervisor swings it in your direction, hitting the back of the chair you are sitting in. What do you do?

The candidate answers are:

A. Swing back at him
B. State unequivocally that such behavior is unacceptable in business and advise him you intend to take this matter up with the manager, to whom you both report.
C. Get up and go straight to the EEO office.
D. Since the boss says, "I was only joking," you ignore the act.

Not surprisingly, response A receives −10. D receives −5 points, with the emphatic comment "Intimidation unchecked is intimidation encouraged. Lack of response will encourage this sort of behavior to expand."

B is given the highest score, +10, with the comment "Not only will this response get the item a third-party inspection, but it will also put your supervisor on notice that you don't accept his action."

The reference to the equal employment office (EEO) in answer C suggests that discrimination may be at work. (In addition to the categories of discrimination mentioned in Title VII, many high-tech companies are also alert to discrimination based on sexual orientation.) The EEO option is scored +5 with the comment "This is your privilege, but it doesn't solve the quality problem." However, the EEO may be most appropriately used if the person suspects discrimination and does not feel comfortable raising that issue with the manager mentioned in B. Companies must be sure their managers can handle issues of discrimination sensitively, if these recommendations are to work.

Mini-case 72 raises some even subtler issues for managers:

When (one) male supervisor talks to any female employee, he always addresses her as "Sweetie." You have overheard him use this term several times. As the supervisor's manager, should you do anything?

The candidate answers are:

A. No, since no one has complained.
B. Yes, talk to the supervisor and explain that, while he may have no sexual intention, his use of "Sweetie" may cause resentment among some of the employees.
C. Yes. Order the supervisor to call an all-hands meeting and apologize for the unintended slights.
D. No, because there is nothing wrong with calling a female employee "Sweetie" or other endearment.

Answer A receives −10 points with the comment, "To some, such informalities are, at best, unwelcome, and, at worst, a form of sexual harassment. Action should be taken to correct the situation even without prodding from an employee." Answer D receives a resounding −20 points and the rebuke "A manager's role is to assure a productive, professional working environment. Option D means you have abdicated."

Answer C receives 0 points and the comment that there is no evidence of harm done without a complaint and so the response is "premature." (Indeed, making such an example of the perpetrator is an overreaction, and a bad, rather than neutral move. That the response is not graded negatively may be compensation for the recent past in which managers tended to be too tolerant of such behavior. Some tendency to go to the other extreme often follows a period of neglect of a problem.)

Answer B receives 10 points with the approving comment "Acting in a firm, nonjudgmental fashion, you are now doing your job as a manager – proactively, not reactively."

The guidance given in these answers clearly shows the concern to go beyond respecting employees' legal rights and promote consideration and good working relationships in the company.

ETHICS IN A GLOBAL CONTEXT

During the development of the *Gray Matters* game the cutoff for the value of gifts that can be accepted from business contacts changed. (The cutoff figure used in commentary on earlier mini-cases was $10. The higher figure of $20 appears in some later ones.) Particular cutoff figures do not carry any moral imperatives behind them, and the lavishness of gifts offered innocently varies significantly from one culture to the next, even from one profession to the next.

U.S. engineers doing business with Japanese firms have often received gifts from those firms that are lavish by U.S. standards. These gifts are normal hospitality by Japanese standards and are given without any expectation that the recipient will do anything improper in return, so they are not bribes. Because they are out of line with what members of U.S. corporations ordinarily give and receive, many U.S. companies doing extensive business with Japanese companies have established practices for dealing with these gifts, such as pooling them and holding an employee drawing for them or giving them to a charity.

Differences in cultural expectations complicate the application of ethical standards across national boundaries, but the formation of new associations such as the European Economic Community and the North American Free Trade Association and the importance of multinational trade are requiring companies to think more cross-culturally about ethics.

At a minimum, U.S. Companies must comply with the so-called Foreign Corrupt Practices Act (FCPA) of 1977. This act makes it a crime for U.S. corporations to accept or offer payments to foreign governments and political parties in order to obtain or retain business. It does not forbid making minor payments to low-level officials to "grease the skids," although the latter might, depending on the situation, also count as a bribe or as extortion.[32] As C. E. Harris has pointed out, there is an important moral difference between bribes and capitulating to extortion in that bribes are paid to obtain something to which one does not have a right, such as a special advantage in awarding a contract.[33] In contrast, extortion is paid to secure something to which one does have a right (or at least a legitimate expectation), such as freedom from arson or the return of equipment one has legally brought into a country but which a corrupt customs official alleges to have been "lost." Although my example of paying extortion for the return of belongings would, other things being equal, be morally justified, paying extortion to a government to prevent them from terminating one's business relationship in which one has made a heavy initial investment may be illegal under the FCPA. Here again we see the distinction between the function of the law and moral evaluation of a particular act. The FCPA was enacted after it came to light that U.S. corporations had paid millions of dollars to foreign governments. The purpose of the law is not only to prevent U.S. corporations from acting reprehensibly, but to discourage extortion by seeing that it does not succeed. To comply with this law and at the same time avoid insulting potential business partners, some companies simply say that the law requires that all payments be a part of any contract.[34] Gifts that are expected as a part of normal courtesy in some culture may be made a part of the contract.

CONCLUSION

In this chapter we have seen that company values influence the relationship between engineers and managers, and engineers' opportunities to fulfill their responsibilities. High-quality organizational complaint procedures may provide means for finding good resolutions of disagreements but are not easy to establish. Establishing the rights, obligations, and responsibilities of employers and professional engineering employees contributes to general understanding. Professional societies, in cooperation with employers, have provided guidelines to establish such norms.

Legal limits on treatment of employees are in force, but these only curtail relatively extreme behaviors. To provide a good working environment that fosters trust and cooperation requires addressing many subtler issues. Farsighted companies and research institutions are concerned with subtler aspects of workplace climate both out of concern for employees and because a good workplace climate fosters high productivity.

NOTES

1. The point is nicely captured in the title of another essay by Michael Davis, "Avoiding the Tragedy of Whistleblowing," *Business & Prof. Ethics Journal*, 8(4):3–19.

2. For a brief comparison of the work situation for engineers in Germany and in the United States, see "Engineering Ethics in the U.S. and Germany" by Adolf J. Schwab in the *IEEE Institute*, June, 1996, which is available on the www at http://www.institute.ieee.org/INST/jun 96/ethics.html.

3. Friedman, Milton. 1970. "The Social Responsibility of Business Is To Increase Its Profits," *The New York Times Magazine* (September 13). Reprinted in *Ethical Issues in Engineering*, edited by Deborah Johnson, Englewood Cliffs, NJ: Prentice-Hall, Inc., 1991, pp. 78–83.

4. Senge, Peter M. 1990. *The Fifth Discipline, The Art & Practice of the Learning Organization*. New York: Doubleday.

5. Michael Davis. 1997. "Better Communications Between Engineers and Managers: Some Ways to Prevent Many Ethically Hard Choices," *Science and Engineering Ethics*, 3(2): in press. (The report is sometimes called the "Hitachi Report," for its sponsor.) The picture of communications at companies that put profits above all else derived largely from the reports of engineers who had formerly worked at such companies and were now working at a quality- or customer-centered company, and so it is less detailed.

6. Michael Davis and his coinvestigators call these companies the "engineer-oriented" companies, "customer-oriented companies," and the "finance-oriented" companies. Quality (along with safety) are values central to engineering, in the sense that an engineer who does not uphold them is seen as a poor engineer; but since quality rather than the engineers (e.g., their happiness or career development) is the focus in these companies, I call them "quality-oriented."

7. *Ibid.* 29.

8. The companies were: Fel-Pro Incorporated, Omni Circuits, Bosch Corporation, W. E. O'Neil, Construction Company, Motorola.

9. Robert Jackall. 1988. *Moral Mazes*. New York: Oxford University Press.

10. Dandekar, Natalie. 1991. "Can Whistleblowing be Fully Legitimated?" *Business and Professional Ethics Journal*, 10(1):89–108.

11. Westin, Alan F. 1988. *Resolving Employment Disputes Without Litigation*. Washington, DC: Bureau of National Affairs.

12. Unger, 1994, p. 226.

13. Eric Pooley, "Nuclear Warriors," *Time*, March 4, 1996, pp. 46–54. Matthew L. Wald, "Two Northeast Utilities Plants Face Shutdown." *New York Times*, March 9, 1996.

14. This scenario was distributed by Doug Heiken in a draft statement, "Making Whistleblowing Obsolete Through Forest Service Reform," dated 12/14/94 and distributed to the large AAAS Science and Engineering Ethics (AAASEST) e-mail list, Thur., 2 Mar., 1995, 09:30:13-0800.

15. Elliston, Keenan, Lockhart, van Schaick, *Whistleblowing: Managing Dissent in the Workplace*, New York: Praeger Scientific, 1985.

16. Schneider, Keith. 1991. "Inquiry Finds Illegal Surveillance of Workers in Nuclear Plants," *New York Times*, August 1, 1991, p. A18.
www A web site on the Hanford Joint Council may be found at http://www.halcyon.com/tomcgap/www/hjc.html. This web site is maintained by the Government Accountability Project (GAP). For a decade, GAP attorneys

have worked with whistle-blowers from the Hanford Nuclear Reservation in Richland, Washington.

17. The current members of the Council are:

    Chairperson: Prof. Jon Brock, University of Washington's Graduate School of Public Affairs;
    Westinghouse Hanford Managers: Richard Slocum, Director of Training Services; Dr. Ronald Lerch, Director of the President's Office;
    Nuclear Safety Public Interest Groups: Gerald Pollet, Executive Director of Heart of America Northwest and Executive Director and Legal Counsel for Legal Advocates for Washington; Tom Carpenter, Director of GAP's west coast office;
    Neutral Members of the Public: Christine Spieth, Secretary-Treasurer of Service Employees International Union, Local 6; Vincent Stevens, former Dean of the School of Health Sciences, Eastern Washington University;
    Whistle-blower Representative: Billie Pirner Garde, former whistle-blower and whistle-blower advocate.

18. See the Hanford Joint Council's web site at http://www.halcyon.com/ tomcgap/www/hjc.html. This web site is maintained by the Government Accountability Project (GAP). The GAP is a public interest group that supports conscientious employees who seek to raise issues on waste, fraud, abuse, threats to public health, and worker safety and environmental hazards. The GAP has been in operation since 1977, and for a decade attorneys have worked with whistle-blowers from the Hanford Nuclear Reservation in Richland, Washington. The GAP reports that whistle-blowers at Hanford had experienced "harassment and retaliation, ranging from management intimidation, security clearance revocation, professional blacklisting, dismissal, and even home break ins."

19. This article, "Resolving Ethical/Technical Dissent Through Due Process," is available on the www at http://www.flsig.org/fcieee/sections/melb/datalink/ sepoct.html#Ethical. A slightly longer version was published in the August 1996 (vol. 6 No. 2) issue of *Engineering Ethics Update*, the newsletter of the National Institute for Engineering Ethics.

20. I happened to witness one bit of a continuing incident when a student working on the WWW Ethics Center first contacted Inez Austin at the Hanford facility about putting her story in the WWW Ethics Center. The letter was returned to us marked "Moved, not forwardable." I then used telephone information to obtain Inez Austin's home phone number in Richland, Washington and found her living in the same house she had occupied for decades.

21. It is also available in the WWW Ethics Center for Engineering and Science ethics.cwru.edu.

22. The foreword to the Guidelines state, "Where differences in interpretation occur, they may be referred to the headquarters office of any of the endorsing societies."

23. The Electronic Frontier Foundation monitors the issue of the privacy of employee and student computer files.

24. See the following works for more on biological testing in the workplace:

    Bird, Stephanie and Jerome Rothenberg. 1988. "To Screen or Not to Screen: Drugs, DNA, AIDS," unpublished manuscript.

Murray, Thomas. 1983. "Warning: Screening Workers for Genetic Risk," *Hastings Center Report* 13(1):5–8.

Ashford, Nicholas A. 1986. "Medical Screening in the Workplace: Legal and Ethical Considerations," *Seminars in Occupational Medicine*, 1(1):67–79.

Rothstein, Mark A. 1987. "Drug Testing in the Workplace: The Challenge to Employment Relations and Employment Law," *Chicago-Kent Law Review* 63:683–743.

Murray, Thomas H. 1992. "The Human Genome Project and Genetic Testing: Ethical Implications." In *The Genome, Ethics, and the Law*, AAAS-ABA National Conference of Lawyers & Scientists. Washington, DC: AAAS.

25. Seth Mnookin. 1986. "Department of Defense DNA Registry Raises Legal, Ethical Issues," *Gene Watch*, 10(1):1, 3,11.

26. Judith A. Lachman, *Issues in Management, Law and Ethics*, Chapter 22, unpublished manuscript. See also *Toward a National Policy on Drug and AIDS Testing*, report of two conferences on drug and AIDS testing, Washington, DC, October 20–21, 1987, and Racine, Wisconsin, March 8–10, 1988, Washington, DC: Brookings Institution, and Walter E. Scanlon. 1980. *Alcoholism and Drug Abuse in the Workplace: Employee Assistance Programs*, New York: Prager.

27. Otis Engineering v. Clark, 668 S.W. 2d 307 (Tex. 1983).

28. The length of time that traces of a given drug or its characteristic metabolites can be found in the urine after use depends on a host of factors, including drug metabolism and half-life, the user's physical condition, fluid intake, and method and frequency of drug use.

www Extensive selections from the Texas Instrument material as well as Lockheed Martin's Gray Matters may be found in the "Ethics In A Corporate Setting" section of the WWW Ethics Center for Engineering and Science. Since advice from the Texas Instruments Ethics Office is not readily available except in the Ethics Center, all references to Texas Instruments advice will be to material contained there.

29. The complete game contained 105 mini-cases. It is now out of print and Lockheed Martin will soon replace it with new materials, also employing mini-cases.

30. Rowe, Mary P. 1990. "Barriers to Equality: The Power of Subtle Discrimination to Maintain Unequal Opportunity," *Employee Responsibilities and Rights Journal*, 3(2):153–163.

31. This scenario forms part of a project that also contains interviews about how best to deal with the issues. It may be found in the Problems section of the WWW Ethics Center for Engineering and Science, http://ethics.cwru.edu.

32. For an NSPE BER case that presents a problem of the sort that prompted this legislation, see case 76-6, Gifts to Foreign Officials.www

33. Harris, Charles E., Michael S. Pritchard, and Michael J. Rabins. 1995. *Engineering Ethics*, p. 108.

34. Lytton, William B. Combating Corruption in Foreign Markets. The Evolving Role of Ethics in Business: Conference Report. The Conference Board, Inc., 1996.

6

RESPONSIBILITY FOR RESEARCH
INTEGRITY

GROUP MISCONDUCT REGARDING
LABORATORY PROCEDURE [1]

You are a graduate student working as part of a group on a large project. The results from your group experiments are used for other experimental work. Your faculty supervisor, the principal investigator (PI) for the project, wants you to use a new procedure in your experimental work. She expects the new procedure to yield results that are better suited to the conditions of the other experimental work. The other members of your group do not want to change the procedure they have been using; the new one requires significantly more work. They believe the PI will not notice if the old procedure is used.

You rely on the group for assistance in your own thesis work, but if you go along with the decision to use the old procedure, the quality of the data will most likely be inferior; you will mislead the PI and perhaps the whole scientific community.

You argue for using the new procedure and informing the PI that the work will just have to take longer – information which she is not likely to receive well. The rest of the group is not persuaded.

What should you do and how can you go about it?

DOUBTS ABOUT PUBLISHED RESULTS [2]

You are a computer science graduate student and for two years have been working on an operating system design in Professor Carr's group. Professor Carr has designed a set of novel heuristics for file-system cache maintenance. Carr published performance graphs describing

194

a simulation of a prototype file input/output subsystem in a journal article and included the graphs in the proposal for the group's current grant. The graphs indicate that Carr's heuristic methods will significantly improve file-system cache performance.

You devise a modification to the file system cache heuristics and asked Carr how to run the simulation code to test the modification. Carr replied that the simulation code had not been used in a long time and had been archived to tape. Carr said it was not worth the trouble trying to remember the archived filenames, because the simulation code was very poor and written in a language that does not run in the group's current computing environment. He told you to write a new, up-to-date simulator.

As you worked on the new simulator, you asked Carr how to simulate several classes of events, but Carr claimed not to remember these details of the old simulator. When you finish building a new simulator, your results are considerably worse than those reported in the performance graphs that Carr published.

You now suspect that Carr did not do a previous simulation and made up the numbers in the performance. Some of your own presentations and papers have been based on Carr's performance data.

What can/should you do?

What, if any, ambiguities do you face?

What risks are there in this situation to yourself or others?

INSTRUCTED TO FABRICATE DATA

You are a senior majoring in computer science and have a part-time job working in the cognitive sciences department of your university. Your job is to write and maintain a computer program to record data on the cognitive performance of patients who are research subjects in a study in one of the cognitive sciences labs.

During the semester the research team writes a grant proposal for continuation of the study that you have been working on. They are behind schedule in their project. Not only have no papers been submitted for publication, but the number of patients tested is not high enough to write any paper. Your supervisor instructs you to "generate numbers" that are comparable to the results they previously obtained and says the numbers you have been asked to "generate" (that is, fabricate) will never appear in the literature and that this is just grantsmanship and everyone does it.

What, if anything can and should you, as an undergraduate, do in this situation?

Where in your institution can you go to get unbiased and confidential advice about the options open to you?

The problem situations in these scenarios range from ones in which the situation is still fluid and offers many possibilities for action to those in which the only remaining actions are to assess a wrong that has been done and decide whether or how to report it. The third presents an extreme situation in which a supervisor proposes a serious violation of ethical standards.[3]

RESEARCH MISCONDUCT OR RESEARCH INTEGRITY?

The responsibility for research integrity has several major components: ensuring the integrity of research results; dealing fairly with others, especially by appropriately acknowledging their contributions; and protecting the welfare of research subjects. Other professional responsibilities of research investigators include laboratory safety and protection of health and safety of the public in the conduct of research.

Recent discussion of researchers' responsibility for the integrity of research has most often begun with consideration of serious wrongdoing in research.[4] This negative approach to the subject of responsibility for the integrity of research is in contrast to discussion of engineers' and scientists' other key responsibilities, such as their responsibility for safety. Even when safety discussions begin with the story of a notorious accident, as have some discussions in this book, suggestion of deliberate wrongdoing is rare. Even negligence and recklessness may be absent when causes of the accidents result from unusual circumstances or lack of knowledge available at the time.[5] They may simply be due to innocent mistakes, mistakes that any competent professional might have made because of the state of knowledge at the time. As Petroski observed, "to engineer is human."[6] Learning from mistakes does not require laying blame for those mistakes, although it is sometimes appropriate to do so.

In contrast, honest mistakes in scientific research are controlled through such mechanisms as peer review of scientific work prior to publication and the requirement for replication of results. The prevention of honest errors is not generally considered to be part of the subject matter of research ethics. Mechanisms such as peer review and replication are imperfect,[7] however, and not equally feasible for all types of research.

Some laboratories replicate their own results before publishing them as a control on mistakes and to ensure their own reputation as scrupulous investigators, but replication is not even possible for large-scale clinical studies. Even where replication is possible in principle, it may be difficult to do if the experiment is expensive to run. Controlling both honest errors and deliberate departures from standards of research practice is demanding.

Widespread public and professional discussion of research ethics has occurred only relatively recently, much more recently than discussions of other responsibilities of engineers and scientists. Research ethics came to attention only after some flagrant cases of research misconduct (and the institutional mishandling of those cases) came to light. Some scientists – for example, Daniel Koshland, when he was the editor of the influential publication *Science* – responded to the concern about research misconduct by saying that too much was being made of a very few notorious offenses.[8] It is true that extreme cases have been overemphasized. The research community has often contributed to this overemphasis and paid less attention than other professionals to more common violations of ethical norms in their work.

The focus on serious wrongdoing in research runs the danger of conveying the impression that concern for a high standard of research ethics is merely an attempt to hold the line against deliberate deception, rather than a concern to develop, maintain, and transmit standards of research integrity in a context of increasing complexity in research practice. The literature on research ethics is so heavily focused on research misconduct, however, that misconduct and the terms used to describe it provide a convenient starting point.

THE SEARCH FOR ADEQUATE TERMS

THE SELECTION OF DATA[9]

Chandler, a third-year graduate student at X Tech, goes to a national laboratory to make a series of measurements on a new experimental semiconductor material using an expensive neutron source. When Chandler returns to X Tech and examines the data, eight of the ten data points show general agreement with a newly proposed theory.

During the measurements at the national laboratory, Chandler observed that there were unpredictable power fluctuations. Furthermore, when Chandler discussed the work with another group doing

similar experiments, Chandler learned that the other group had gotten results confirming the theoretical prediction and was writing a
manuscript describing their results.
How should Chandler write up the results for publication?
Should the data from the two suspected runs be included in tests of
statistical significance and why?

The subject of research misconduct is not entirely new among researchers, though it is new as a subject of public discussion. As far back
as 1830, the English mathematician Charles Babbage wrote an influential book on dishonesty in research.[10] (Babbage was the inventor of
both the "difference engine," and the "analytic engine," the precursor
of the modern electronic computer. He founded several professional
societies.) In his book on research dishonesty he defined several terms
to describe research misconduct, including one that is still very much
in use: "cooking the data." To cook the data is to select only those data
that fit one's hypothesis and to discard those that do not. Selecting data
solely because they support one's hypothesis is misconduct, now called
"falsification" of data. "Cooking," however, is a term that investigators
often apply, sometimes in jest, when describing the use of methods for
data selection that are ad hoc or not fully understood, but which make
"messy" or "noisy" data look more conclusive.

"Data selection," when made according to legitimate criteria, is an
indispensable part of science. It is legitimate to discard some data if that
"run" or sample is contaminated (for example, because you dropped the
sample on the floor) or if statistical methods applicable to the data you
have collected warrant discarding some "outliers." *An essential feature of
legitimate data selection is disclosure of the methods of selection.*

Despite Babbage's early work, the scientific disciplines are having
to reinvent the language to discuss responsible research behavior and
departures from it. The need for new terms is partly due to a period of
silence on the subject but also due to rapid growth and change in research. Data now come in many forms, no longer just observations to be
recorded in laboratory notebooks, but, for example, photographs and
micrographs as well. Using a bound notebook with numbered pages for
laboratory observations is not possible with photographs or computer
printouts. As science and engineering have become more specialized,
the need for collaboration among investigators with different expertise
has increased. Researchers often have only a very general idea of the

standards of research practice that apply to the work of their colleagues in other disciplines. The need for large-scale studies has produced new collaborative arrangements among many individuals and often across many institutions; this can create new occasion for error, confusion, and misrepresentations[11] and has led to questionable findings and eroded public confidence in the value of research.

A set of terms adequate to express morally relevant features of moral life are much needed. Clarity is needed about "research conduct" itself. "Scientific conduct" is sometimes used as a synonym for research conduct to the confusion of all who are aware of how much more there is to science than research. For example, the authoritative introduction to research for students is titled *On Being A Scientist*. The National Academy of Sciences committee that authored the first edition was called the "Committee on the Conduct of Science." [The authoring committee for the second edition is Committee on Science, Engineering and Public Policy (COSEPUP), a committee of the National Research Council (NRC), the research arm of all three national academies – the National Academy of Sciences (NAS), National Academy of Engineering (NAE), and Institute of Medicine (IOM).[12]] A convocation on research ethics the NAS held in June 1994 was called a "Convocation on Scientific Conduct" much to the confusion of some attendees who expected an agenda broader than research.

The misleading breadth of the term "scientific conduct" may have gone unnoticed because many scientists of previous generations were slow to recognize any socially and ethically significant results of scientific work other than the growth of knowledge through research. Work in the history and philosophy of science has discredited the idea that the methods and judgments of science are value-free, although the place of values in science is not the same as the place of values in, say, politics. Authoritative statements from science and engineering bodies now reject the myth of value-free science. For example, although quite different in most respects, both the original (1989) and the second (1995) edition of *On Being a Scientist* recognize the place of values in science. They discuss both the values involved in judging what differentiates a good explanation or theory from a poor one – called the values "internal to science," such as simplicity, consistency, and the ability to yield accurate predictions – and the values that an investigator carries over from other aspects of life into scientific work – the values "external to science." (In terms of the types of value in the introduction, the values internal to science are primarily epistemic values.)

In summary: That "scientific conduct" is still commonly used as a synonym for "research conduct" reflects the recent character of awareness by scientists of the place of values in their work and awareness of themselves as professionals whose actions affect the public good in multiple ways and the lack of an established vocabulary with which to discuss researchers' responsibilities.

WHAT COUNTS AS RESEARCH MISCONDUCT?

The term "research misconduct" is not applied to all types of wrongdoing that occur in a research setting. For example, if an investigator takes home pieces of lab equipment for personal use, that would count as stealing, or at least as misappropriation of property, rather than as research misconduct.

One common attempt to explain this exclusion is to say that "research misconduct" covers only wrongdoing unique to the conduct of scientific research. If "research misconduct" covered only wrongdoing unique to research, it would exclude plagiarism.

Failure to follow standards of good research practice also does not itself qualify as research misconduct, even if it jeopardizes research results. For example, deciding not to run adequate controls certainly departs from standards of good research practice. Some departures from good research practice are more than "honest mistakes"; they are signs of incompetence. But the ethical responsibilities of investigators do not include the responsibility not to undertake work outside their competence, whereas engineering codes and guidelines emphasize this strongly.

More surprising, abuse of human or animal research subjects is not usually counted as research misconduct in the United States, although such behavior can result in the same penalty of debarment from future grant competition. This exclusion results from the linking of the definition of misconduct to jurisdictional decisions over who should investigate what charges. Since the abuse of experimental subjects was already regulated, it did not need to be included under new "misconduct" regulations. The term, "research misconduct" should, therefore, be regarded as what philosophers call a *technical term*, that is, a term defined to serve a special purpose and whose meaning is not simply what you would expect from the meanings of the words in ordinary conversation. For example, "power" and "force" are technical terms in mechanics because they do not carrying the same meaning as in ordinary use.

In the United States, definitions of misconduct proposed in the 1980s began by specifying three acts that exemplify[13] or constitute research misconduct: **fabrication** – making up data or experiments or other significant information in proposing, conducting, or reporting research; **falsification** – changing or misrepresenting data or experiments or other significant information such as the investigator's qualifications and credentials; and **plagiarism** – representing the work or ideas of another person as one's own.

The definition of "misconduct in science and engineering" that the National Science Foundation (NSF) uses is:

> fabrication, falsification, plagiarism, or other serious deviation from accepted practices in proposing, carrying out, or reporting results from activities funded by NSF; or retaliation of any kind against a person who reported or provided information about suspected or alleged misconduct and who has not acted in bad faith.

The NSF and the Department of Health and Human Services (HHS), which includes the National Institutes of Health (NIH), are the agencies with the greatest regulatory oversight over research. Therefore, their definitions determine the conditions under which charges of wrongdoing are the business of those agencies.

The NSF[14] and the HHS continue to clash with organizations such as the national academies and some professional organizations and universities to which researchers belong over the definition of misconduct. This struggle plays out in disputes that appear to be about definitions, with research organizations objecting to the phrase "or other serious deviation from accepted practices in proposing, carrying out, or reporting results."

The NRC, the research arm of the national academies, convened a panel to look into ethical standards of research conduct and they issued their report, *Responsible Science*, in 1992. That report defines misconduct as "fabrication, falsification, and plagiarism in proposing, conducting, and reporting research," and some universities followed suit.[15] The panel had strenuously objected to the inclusion of "other serious deviation from accepted practices," on the ground that it might lead to the filing of a complaint of misconduct against investigators "based solely on their use of novel or unorthodox research methods."[16] The NRC panel classed as "other misconduct" much of what the NSF counts as "other serious deviation from accepted practices," that is, something that is clearly wrong but most of which is not business of the government overseers to control. They did recognize, however, that some of what they call "other misconduct," in particular retaliation against a whistle-blower, is *directly associated* with research misconduct.

The dispute about the definition of research misconduct really centers on the latitude that the government should have in deciding what actions to investigate rather than appropriate concepts for understanding research ethics. In an important sense the universities and other organizations speaking for investigators only object to the breadth in the definition the government uses, since the policies and procedures manuals of many research universities show even those universities that object to latitude in the NSF definitions broadly define the wrongdoing that they will consider. [17] As we saw in Chapter 3, professions generally seek autonomy, so it is not surprising that the research community is wary of government regulation. To be relatively free of such regulation, the research community needs to further develop its own ability to regulate itself. It is in everyone's interest that it do so, but such self-regulation is not yet well developed. It had been hoped that normal mechanisms of research practice would accomplish the task, but studies have shown that such mechanisms are not sufficient even to purge bogus results from the literature in a reasonable amount of time. [18]

In the meantime, some of the actions that the NSF or HHS have counted, or said they would count, as research misconduct under the "other serious deviation" clause include cases of sabotage of experiments (as contrasted with general vandalism of a research site) and setting up a training program as a means for coercing sexual favors from trainees (as contrasted with sexual harassment in general). [19] Furthermore, funding agencies explicitly include deception in proposing research in their definition of misconduct. They have considered lying in a grant proposal about something other than research results to be "other serious deviation from accepted practice." Presumably, this is because "fabrication and falsification" are reserved for misrepresentation of research results and methods. This point is illustrated in a recent decision by the NSF Inspector General's Office – their first finding of misconduct against *an institution*. [20] In that case, two principal investigators at a college of engineering in the southwest submitted three proposals and two letters to the NSF that contained false statements. These statements grossly exaggerated the services offered to Native American and Hispanic students at that college of engineering. (These proposals were submitted to NSF education programs. These programs emphasize service to minorities that are underrepresented in science and engineering, so the false statements greatly strengthened the proposals.)

The Office of the Inspector General considered this to be misconduct and recommended to the Director of NSF[21] that both the PIs and the college of engineering be found guilty of misconduct. They took the position that institutions must "take responsibility for the truth of statements in proposals that concern matters within the purview of the institution itself." It is hard to object to government funding agencies pursuing such malfeasance under *some* authorization, whatever it is called, but the cases do differ significantly from one another.

In the fall of 1995 the Commission on Research Integrity, set up by the HHS, produced a report with a new and much longer definition of misconduct, this time including examples. The drafters avoided specifying the acts that count as research misconduct: Their definition discusses general criteria for judging some wrongdoing to be research misconduct. They say, in part,

> research misconduct is significant misbehavior that fails to respect the intellectual contributions or property of others, that intentionally impedes the progress of research or that risks corrupting the scientific record or compromising the integrity of scientific practices.

This definition also has its critics.[22] Definitions of misconduct are likely to move away from the simple "falsification, fabrication, and plagiarism" mold but to remain controversial for the immediate future.

DOES "SCIENTIFIC FRAUD" DESCRIBE RESEARCH MISCONDUCT?

The greatest attention has been given to cases of research misconduct that are quite extreme. In one rather bizarre case, William Summerlin, while researching transplantation immunology, engaged in several acts intended to deceive others into believing that it is possible to lessen transplant rejection by placing donor organs in tissue culture for a period before transplanting them. This had been his original hypothesis; but when experiments failed to confirm it, he engaged in several crude fabrications, one of which was to use his felt-tip pen to darken the fur patches that he had grafted from black mice to white mice.[23] He also exhibited some rabbits that he reported had received human corneal transplants in each eye, one being a fresh transplant that was unsuccessful and the other a cultured transplant that was successful. In fact, the rabbits had each received only one cornea, and all transplants were unsuccessful. Summerlin's misconduct came to attention after a laboratory assistant reported his act of painting the mice to the director of the Sloan-Kettering Institute.

The image of painting the mouse was incorporated into the 1988 Nova video *Do Scientists Cheat?*[24] and became a key image for research misconduct in the public imagination. This video helped establish the expression "painting the mice" for flagrant acts of misconduct. This colorful phrase and the Summerlin case have contributed to a misleading picture of research misconduct as crude fraud. Even the first (1989) edition of *On Being a Scientist* used the Summerlin case to introduce their discussion of what they called "fraud in science."[25]

The term "fraud" was once widely used to describe research misconduct. In 1985 the U.S. Congress passed a section of the Public Health Service (PHS) Act titled "Protection Against Scientific Fraud."[26] The *Do Scientists Cheat?* video presents statements from many knowledgeable figures who liberally used the term "fraud." The term continues to be used by some knowledgeable writers,[27] but is now primarily used in the popular press and in informal discussions. The National Research Council's Panel on Scientific Responsibility and the Conduct of Research, in *Responsible Science*, argued that "fraud" should not be used to describe research misconduct. They said:

> [M]ost legal interpretations of the term "fraud" require evidence not only of intentional deception but also of injury or damage to victims. Proof of fraud in common law requires documentation of damage incurred by victims who relied on fabricated or falsified research studies.[28]

David Goodstein, vice provost at California Institute of Technology from 1987 and the chief architect of Caltech's regulations on "scientific fraud," made similar observations in a 1991 article.[29] The current (1995) edition of *On Being a Scientist* does not use the term.

Aside from the requirement that there be a party who has been injured by the fraud, the legal notion of fraud has three basic elements:

1. the perpetrator makes a false representation;
2. the perpetrator knows the representation is false or recklessly disregards whether it is true or false; and
3. the perpetrator intends to deceive others into believing the representation.

Do most instances of misconduct meet these three criteria? The 1989 edition of *On Being a Scientist* claimed that "[t]he acid test of fraud is the intention to deceive." Is most misconduct characterized by an intent to deceive others into believing true something one knows to be false or at least that one has no good reason for thinking true?

FABRICATION: FROM HOAXES TO UNDONE WORK

Famous hoaxes, such as the Piltdown man hoax, are included in both Broad and Wade's *Betrayers of the Truth* and in the *Do Scientists Cheat?* video. The term "hoax" has the connotation of fooling others for the sake of doing so rather than for some other end such as appearing to be a productive researcher. Perpetrators of hoaxes are often anonymous. "Piltdown man" was a hoax, created by passing off a combination of human and ape bones as the remains of a single humanoid "missing link." Such famous hoaxes in science are intentional deceptions, but these are quite different from most cases of research misconduct. Hoaxes are rare in the natural sciences, medicine, and engineering. The few hoaxes that I have found in these areas have been hastily concocted ruses, such as rumors about computer viruses, "Trojan horses," and "e-mail bombs."

The story of Cyril Burt's fabrication of experiments about the heritability of intelligence is another case of a deliberate deception that appears in *Betrayers of the Truth* and in *Do Scientists Cheat?* Cyril Burt's supposed study of the intelligence of identical twins who had been separated at birth is one of those often-cited cases of fabrication that does meet the three criteria for fraud discussed above, namely, that there be intentional deception in which the perpetrator seeks to induce others into believing something that is false and that the perpetrator knows to be false or at least has no reason to think true. Most of Burt's research subjects were fictitious. His fabricated research furthered the belief that inheritance overshadows environment in the development of intellectual ability, and it led to the adoption of educational policies built on that belief. Burt may have deeply held this belief, but he knew that neither the majority of the children he supposedly studied nor his two research assistants existed.[30] He also knew that he had no basis for his quantification of the supposed predominance of heredity over environment. However strong his preconceptions, Burt disregarded the accuracy of his figures. Indeed, it was the invariance in his reported quantification of the influence of heredity over environment in three separate studies that led Leon Kamin, the person who first exposed Burt, to suspect fraud.[31] Burt's quantification is certainly false, and he clearly sought to deceive others into thinking it true. Therefore, Burt's actions fulfill the criterion that he led others to believe what he knew to be false.

Fulfillment of this criterion is rare in the natural sciences, even in cases where misconduct is found, because one knows detection of misconduct is more likely if one fakes evidence for a *false* conclusion. A more common scenario is: The perpetrator believes in the truth of a certain conclusion and has at least some grounds for that belief but acts to deceive others about the nature and strength of the evidence for that conclusion. In this respect, James Urban's fabrication of research results better exemplifies typical fabrication in contemporary research.

James Urban was a post-doctoral fellow at Caltech who was found to have fabricated data in a manuscript he submitted to the journal *Cell.* He claimed that the data reported in the published version of the paper were genuine. They were certainly different from those in the manuscript originally submitted to *Cell.* Some of Urban's lab books were missing and so could not be examined. (He said that they were lost in a subsequent move across the country.)

Urban did not deny the charge of fabrication, but he did deny any *intent to deceive.* Clearly he did intend to lead the reviewers for *Cell* to think that he had obtained experimental results that he had not in fact obtained; that much intent to deceive is implied by the term "fabrication." One official close to the case said that Urban believed he knew how the experiment would turn out and, because of the pressure to publish, tried to "speed" the review process by fabricating the data in the original manuscript. But the official was convinced that Urban would not have published without having first inserted data he had actually obtained experimentally.[32] So the point of Urban's denying an intent to deceive was that he did not intend to deceive others about natural phenomena, that is, commit fraud in the strict sense. Apparently, Caltech also understood Urban's actions in this way because they found him guilty of "serious misconduct" but not of "fraud," which Caltech distinguished from "serious misconduct" and regarded as a graver charge. The *Cell* article was retracted.

Another realistic example of fabrication of data appears in a scenario at the beginning of this chapter about fabricating data for a grant proposal. These examples illustrate that fabrication of research results, although a serious misrepresentation, often does not represent as true a research conclusion that the perpetrator knows to be false or has no sound basis for thinking true. Although each case is unique, what we commonly find is a situation in which an investigator has some scientific evidence for a conclusion but seeks to deceive others about the strength of that evidence. All these cases are quite unlike the Summerlin case.

Fabrication of *plausible* findings, no less than fraud in the sense of representing as true what one knows to be false, thwarts the progress of scientific research and confuses the application of research results for human benefit. Where the accusation of intent to deceive is raised, it is important to ask whom the perpetrator is intending to deceive about what, since although all deception may be wrong, or at least need moral justification, deceptions differ in ethically important respects.

THE RARE CASES OF FRAUD

In 1989, Efraim Racker published an article in *Nature* entitled "A View of Misconduct in Science."[33] It gives a detailed account of the case of a sophisticated deceiver, one who was much more clever and calculating than Summerlin or Burt. The perpetrator was Mark Spector, who at the time of his deceptions was Racker's graduate student. For an extended period, Spector deceptively altered reagents to produce experimental results that made it appear he had impressive experimental verification for Racker's new theory of cancer causation, known as the "kinase cascade" theory. Racker describes Spector as a brilliant researcher and gifted teacher with multiple talents, a person with such obvious gifts that he had no need to secure his position through deception. That he did perpetrate this elaborate deception is attributed by Racker to mental disturbance. (Spector was later found never to have obtained his bachelor's or master's degree and to have pleaded guilty of forging checks before coming to Cornell.) Racker is quick to clarify that deceptive researchers like Mark Spector are not psychotic, even if mentally disturbed, and that they should be prosecuted for their misconduct.

The examples of Burt, Spector, and three other calculating perpetrators of fraud in the natural sciences whom Racker mentions warn us against the error of claiming that perpetrators of misconduct do not mean to mislead people about natural phenomena. Spector and the other perpetrators of fraud, with the possible exception of Burt, may have started by faking data in support of what they believed were, and would be shown to be, true conclusions. Only after they were in over their heads did they continue to fake data in support of conclusions in which they had no confidence, and thus commit fraud. The 1996 case of fabrication and falsification by a graduate student of NIH director of Genome Research Francis Collins, appears to have begun when the student had difficulty creating a control cell line for a test and made up the data he thought he should be able to get. Later he went on to recklessly faking results that Collins thinks he should have known would be discovered.[34]

Acts of falsification and fabrication themselves entail at least an intent to deceive others about one's data, but, as the Urban case illustrated, that deception is more likely to be about the strength of the evidence for a given conclusion rather than the character of the conclusion itself.

DISTINGUISHING FALSIFICATION FROM LEGITIMATE DATA SELECTION: ROBERT MILLIKAN'S DATA HANDLING

To falsify experimental data is to change the data to fit the investigator's expectations or preferences. Falsification might take the form of altering the recorded value of particular bits of data observation, or it might take the form of "cooking." Changing the value of some data is absolutely prohibited, so no question arises of when it is justified to do it. However, excluding some data points or smoothing the curve plotted from the data may be justified data selection or unjustified "cooking." What is the difference?

The selection and presentation of data are a professional responsibility and require the exercise of judgment. Discretion is required to recognize sources of "noise" (that is, extraneous influences on observations of the phenomena under investigation) and to apply statistical methods to deal with noisy data, even where the source of the noise is unknown. Making the required judgments is therefore more complex than simply reliably recording data. Self-deception is also more of a risk when one must exercise discretion.

The complexities involved in data selection are well illustrated in the story of Robert Millikan, who in 1923 won the Nobel prize for his work establishing that the electron carries a characteristic amount of charge rather than varying in the charge it carries. The story of his data selection in that work is interesting for its bearing on both scientific method and research ethics. It provides a historically interesting example of the place of intuition in science and of the evolution of physics in the early twentieth century.

Intuition is the ability to immediately recognize what is going on in a situation. There need not be anything mysterious about intuition; it is often just what is apparent to "the experienced eye," but not to every eye, or ear, or touch. In contrast, the ability to infer what is going on from other, independently identified evidence or premises, is called *reasoning*. The ability to recognize something without being able to articulate the basis for one's recognition is familiar in everyday life. One may recognize an acquaintance at a great distance just from the person's walk, without being able to say what it is that is distinctive about that walk. Parents can regularly distinguish the cry of their own child from that of other children, although few are able to describe what is distinctive about the cry of their own child.

Describing some of Millikan's data selection as the operation of his "intuition" rather than as "reasoning" suggests that Millikan could not articulate exactly what feature of his observations made him think that something was amiss with some of his experimental observations. He was often able to think of reasonable explanations of why "things went wrong" with the experimental situation, however. Among the hypotheses that he offered for what "went wrong" in some experimental runs were that "two drops stuck together" or that "dust" interfered. Some of these hypotheses helped him improve conditions in subsequent experimental preparations.[35]

In practical areas of scientific work, such as engineering practice or clinical medicine, one's professional standing is based on the successful outcome of their practice – for example, the high quality, safety, and reliability of the products a person has designed or manufactured or the good health outcomes secured for patients – rather than solely on the quality of the reasons a person can give for professional decisions. In most areas of professional practice, reliance on intuition or professional experience is well accepted. It is because of the importance of developing the practitioner's "experienced eye" or intuition that education in practical scientific areas includes internship or apprenticeship experience that allows the student to "get a feel for" what is important and how experienced professionals go about their work. Of course, if a practitioner turns out to be wrong very often, others stop seeking his professional opinion.

Scientific research has a much more difficult time dealing with judgments based on intuition and experience, as contrasted with those based on reasons that can be fully articulated. This is due in large part to the role of peer evaluation in deciding which research results are worthy of publication. In evaluating the results of an experiment, investigators' reasoning, and not just their conclusions, are subject to scrutiny. Saying "I discarded all the data taken when there was something funny going on in the experiment" may describe the operation of true insight, but it is not likely to be very convincing to reviewers. Reliance on intuition leaves one especially vulnerable to self-deception, but research cannot dispense with intuition either, as the case of Robert Millikan illustrates.

The full story of Millikan's research makes fascinating reading. It has been thoroughly researched and engagingly written by science historian Gerald Holton in two articles.[36] Holton convincingly argues that Millikan's intuition was part of what made him a better researcher than his rival, Felix Ehrenhaft. In particular, Millikan's ability to recognize and select the data that most accurately reflected the underlying phenomenon was what put him ahead of Ehrenhaft, who indiscriminately used all of his data and therefore came to the wrong conclusion.

As those who have repeated Millikan's oil drop experiment know, it is often evident that "something funny" is going on – for example, two oil drops may stick together and behave in ways that are different from the behavior of a single oil drop. If an investigator has independent basis for believing that some data are flawed (a basis other than that the values obtained are not the ones expected), then the investigator has some justification for excluding those data points. Of course it is important to be even-handed in excluding data that are suspect, discarding both those that do and do not support one's hypotheses. For example, if one of the instruments is discovered to be malfunctioning and the investigator discarded all data back to the last time the instrument was tested and found to be in good working order, that would be methodologically and ethically justified. Of course, that might also mean discarding a great deal of data.

A present-day evaluation of Millikan's work is complicated by the fact that today's standards for data selection were just developing in Millikan's time. Indeed, Millikan helped develop them. The evidence that he was operating under different methodological criteria is part of what makes the story of his research significant for the history of science and the history of research methodology in particular. For example, Holton quotes passages from Millikan's 1910 paper in which Millikan frankly states views, methods, and attitudes toward data handling that sound outlandish by contemporary standards:

> in the section entitled 'Results,' Millikan frankly begins by confessing to having eliminated all observations on seven of the water drops. . . A typical comment of his, on three of the drops, was: 'Although all of these observations gave values of e within 2 percent of the final mean, the uncertainties of the observations were such that I would have discarded them had they not agreed with the results of the other observations, and consequently I felt obliged to discard them as it was.' Today one would not treat data thus, and one would surely not speak about such a curious procedure so openly.[37]

That Millikan was so open about the methods he used demonstrates how far he was, in 1910, from attempting to deceive anyone. Furthermore, that his paper containing these comments was published in a very prestigious journal shows that this description of his data handling did not strike his contemporaries as very odd. However, present-day standards were indeed developing. Three years later, when Millikan published a major paper on the character of electronic charge based on his oil drop experiment, he seems to have become self-conscious about the issue of selecting data. In this 1913 paper he writes, in italics,

> *It is to be remarked, too, that this is not a selected group of drops but represents all of the drops experimented on during 60 consecutive days.*[38]

Regrettably, Millikan's statement is false; he certainly knew it was when he made it. It is instructive to examine both this moral lapse on Millikan's part and the heated debate and curious silences about this lapse that have occurred within the scientific community since 1981.

Granted that Millikan was wrong to lie, why did he do it? There is no evidence that Millikan lied about other matters. Indeed, the passage quoted above from Holton shows a praiseworthy openness about his methods (even if the methods themselves look peculiar by our standards). A plausible interpretation is that in the years between 1910 and 1913, Millikan had become aware of the emerging opinion that researchers ought to give reasons for discarding data. Perhaps someone challenged him on the passage cited above from his 1910 paper. In the research for his 1913 paper, Millikan had used methods for data selection that he could not fully explain. By today's standards, Millikan would have been expected to give reasons for discarding some readings. By the standards of his own time, however, it would have been acceptable if he had published his data without comment on his data selection; that is, by the standards of his time his data handling was not misconduct. His only failing was his gratuitous misrepresentation of his method.

THE DEBATE AND THE SILENCE ABOUT MILLIKAN'S LIE

Millikan's act has been evaluated very differently by a variety of commentators since Gerald Holton first examined Millikan's notebooks and brought to light the discrepancy between Millikan's data selection and what, in his 1913 paper, he said he did. The discussion reveals much about the present anxiety and confusion that surround many discussions about research ethics.

Holton himself did not comment extensively about Millikan's moral lapse, but rather he was content to display the discrepancy in detail, not only in quotations from Millikan's work but in a reproduction of two pages from Millikan's laboratory notebook dated March 15, 1912. Among the clearly written comments are "Beauty" on the left-hand page by the data taken from one drop and, on the right hand page, "Error high will not use" by another drop. What we do not and cannot know is the extent to which Millikan's data selection was influenced primarily by noticing "something funny" in the experimental behavior of the drop other than that it was behaving in a way that would not yield the value of electron charge that Millikan expected. This means that we do not know how purely Millikan's intuition was operating. Given what we now know about the role of "observer bias" in experimental observation (roughly that people tend to see what they expect to see), this may be something that Millikan himself could not have known.

Another paper, "Millikan's Published and Unpublished Data on Oil Drops,"[39] which was published two years after Holton's and in the same journal, made much more of the discrepancy between what Millikan had done and what he claimed to have done. This latter paper was then drawn on by William Broad and Nicholas Wade for their book on research "fraud," *Betrayers of the Truth.*[40] Subsequently, the booklet *Honor in Science* was written by C. Ian Jackson for Sigma Xi, drawing on all three of the other sources.

Honor in Science was groundbreaking as a statement on research ethics from a society of researchers. Since the claim of any professional group must be self-regulating if it is to be free of regulation by others, this was an important initiative by Sigma Xi, even though as an honor society it has few sanctions it can apply other than exclusion from its own membership.

Honor in Science calls Millikan's act "cooking the data." However, if, or insofar as Millikan had criteria other than the value of e obtained from a particular experimental run as a reason for rejecting the data from that run, then he did not "cook" the data, but lied about omitting those drops he saw as having behaved oddly.

The 1989 edition of *On Being a Scientist* included discussion of the fact that Millikan's data selection methods would now be unacceptable, together with a color reproduction of the same two pages from Millikan's laboratory notebook for March 15, 1912 that appear in Holton's 1978 article. However, it *omitted any mention of what Millikan had said about his data selection.* Not surprisingly, in a discipline that prides itself on striving for impartial evaluation of evidence, the omission of any mention of Millikan's lie, juxtaposed with the account of that lie in *Honor in Science*, led some scientists to cynically conclude that *On Being a Scientist* was cooking the data by omitting evidence of wrongdoing by a famous scientist. (The 1995 edition of *On Being a Scientist* excludes mention of Millikan. This action, too, occasioned some acid comments.) The now familiar polarization on the question of Millikan's behavior gives further evidence of the current difficulty in recognizing and discussing any abuses or moral lapses other than research misconduct, especially where a famous scientist is concerned.

RESPONDING TO CHANGING MORAL STANDARDS

The circumstances that Millikan faced are not unusual, and they warn us to pay attention to the moral problems that arise when standards of good practice change. Human standards are always imperfect and subject to revision. It is a source of both excitement and frustration in rapidly developing areas of engineering, science, and medicine that some standards can be expected to change during one's professional career. Chapter 3 discussed in some detail the changes in the scope of matters that an engineer is expected to consider to ensure safety. Henry Petroski's book, *To Engineer is Human*, deals extensively with the topic of how engineers learn from accidents. Similarly, the standards for environmental protection in general and for waste disposal in particular have changed radically during the professional life of many practicing engineers. Clinical medicine changes continually, so the preferred treatment in one decade may become malpractice in the next. The ethical standards for the use of human subjects in research have changed substantially since World War II.

Conscientious engineers have said that at the beginning of their careers they had often disposed of some hazardous substance by "throwing it out the back door" (that is, disposing of it carelessly without considering where it would end up). Deans of medical schools have volunteered that when they started out, the rule they used for an ethically acceptable experimentation on another person was that they had first done the procedure on themselves. They never thought about consent. They were responsible practitioners, but they saw the standards by which they once practiced criticized and superseded by more stringent standards. These engineers and physicians often seemed somewhat embarrassed by the discrepancy between their past acts and present standards, but they were honest with themselves and others about what they had done.

Engineers and scientists who are beginning their careers can expect to encounter changes both in the standards of ethics and in what is demanded as a part of competence in their fields. Therefore, the challenge of coping when the standards that were previously met are no longer morally or technically acceptable is one that any engineer or scientist should be prepared to meet. Millikan probably had never thought about the possibility of a change of standard. When faced with the problem of how to cope with one, he told a gratuitous lie that has become an ugly little blotch on an otherwise stellar scientific career.

FROM HONEST MISTAKES TO NEGLIGENCE
AND RECKLESSNESS

Often a person or organization accused of wrongdoing will plead that their action was only a mistake, an oversight due to carelessness, but that there was no intention to deceive. Awareness of this tendency may explain the willingness of some to refer to research misconduct as scientific "fraud" and to regard the intention to deceive as the acid test of fraud.

As we saw at the beginning of the discussion of misconduct, the Committee on the Conduct of Science, in the 1989 edition of *On Being a Scientist*, took this position and offered a simple contrast between intentional misconduct on the one hand and mistakes or errors on the other.[41] They argue that honest mistakes are a necessary byproduct of scientific progress. The NAS panel, writing three years later, observed that although intent is important,

> especially in the adjudication of allegations of misconduct, intention is often hard to establish and does not provide, by itself, an adequate basis for separating actions that seriously damage the integrity of the research process from questionable research practices or other misconduct.[42]

In Chapter 2 we examined the characteristics of mistakes that contribute to making them innocent or blameworthy and excusable or inexcusable. Those distinctions are fully applicable to the assessments of mistakes in research practice. Although question of the intent to deceive commonly arises in assessments of questionable research conduct, intent to deceive is only rarely an issue in assessing whether behavior failed to fulfill the standards of responsibility in engineering practice. For example, when Jack Gillum affixed his seal to the unsafe modification that the fabricator sought to make on plans for a connection in the ill-fated Kansas City Hyatt Regency, there was no question of his having any nefarious intentions; he seemed to have approved the changes without giving adequate attention to what he was approving. Competence and care are elements of professional responsibility. Failure to give adequate attention and care more often than evil intent leads to a failure in professional responsibility.

One important difference between research and the design, development, and manufacture or construction of devices, processes, and structures is that investigators receive no independent feedback analogous to the performance, accident, or failure that engineers observe in their devices, processes, or structures. Decades often pass before an investigators' account of some phenomena finds independent confirmation. Adherence to good research practice is, therefore, the sole source of assurance for research investigators.

Researchers may be simply mistaken in either trivial or serious respects. Trivial careless mistakes, such as typing errors and minor omissions, are correctable through errata statements. Even serious mistakes, ones that invalidate an entire work, need not be ethical failings, however. Serious mistakes may be made because the phenomena investigated are beyond the capacity of science to understand at this stage, or because the investigators lack the technical competence to conduct the research undertaken. As long as the investigators have not misrepresented their technical qualifications to do the research, lack of technical competence is not regarded as a moral failing in research.

In contrast to research, practicing beyond one's competence in professions such as engineering, medicine, or law is recognized to be irresponsible. In those professions, grave harm to others may result from incompetent practice, so they recognize an ethical responsibility to practice only within the limits of one's competence. Although the grave health and safety threats from some research accidents (such as the 1979 outbreak of anthrax from a secret military microbiology facility in Sverdlovsk in the former Soviet Union) have led to widespread blame of investigators involved for irresponsible behavior, incompetence that leads to mistaken results rarely has consequences beyond wasting resources, because such results infrequently receive much attention, if they are published at all.

We saw in Chapter 2 that "negligence" is a term of moral judgment and that some mistakes – negligent or reckless mistakes – are morally blameworthy.[43] A careless act shows insufficient care and attention; a negligent one shows insufficient care in a matter where one is morally obliged to be careful. Reckless (or "grossly negligent") acts in professional practice ignore dangers that should be obvious to a minimally competent professional so the acts themselves create a presumption of willfully ignoring those dangers together with failing to give them due attention and care.

Some purported mistakes are so outrageous that it is hard to believe they are simply mistakes. For example, Stanley Pons and Martin Fleischmann sent in an errata statement (which is usually a list of typographical errors and the like, submitted after an article is published) regarding their initial publication on cold fusion, which included the name of an additional author, graduate student Marvin Hawkins.[44] The omission of an author is not an action that researchers, even careless ones, ordinarily make, and in fact there was more to the story of the omission of Marvin Hawkins's name. Hawkins was a graduate student who worked on a large part of the research on cold fusion with Pons and Fleischmann. Just before Pons and Fleischmann announced that they had achieved cold fusion in March of 1989, Hawkins was informed that his name would not be included in the list of authors, because the University of Utah wanted him to "play a lower-key role."[45] The situation escalated when Pons accused Hawkins of stealing the lab books. Hawkins claimed that Pons and Fleischmann were aware, and had suggested, that he put the notebooks in a safety deposit box. The books were eventually returned, but Hawkins was replaced on the project and asked by Pons to leave the University. With Fleischmann's help, Hawkins returned to working on his thesis, and he was added to the list of authors in the errata.[46]

Being overtired or in a rush to meet a deadline are often given as explanations for negligence in performing tasks or fulfilling responsibilities. The degree of culpability for instances of negligence depends both on the egregiousness of the act and on the extent to which the factors that compromise performance are under the agent's control.

Part of the current concern about research ethics has been precipitated by the observation that some reputable researchers were behaving in grossly negligent ways and by their negligence calling into question the quality control of scientific research.

In nonresearch contexts, describing an act as "cutting corners" may mean just taking some shortcut, but in connection with research practice the phrase is regularly used to describe practices that, unlike honest mistakes or even careless mistakes, are knowingly undertaken violations of the standards of good research practice. Even though it is not the researchers' intention to put bogus results into the literature or to misappropriate credit, their actions run the risk of doing so. Harvard Medical School's Eleanor Shore, the dean charged with oversight of research ethics, observes that personal expediency is often the motive for such acts.[47] Expediency leads perpetrators to recklessly disregard accepted standards of research practice and so puts at risk the integrity of research results or proper assignment of credit for research.

Notice that some acts violate accepted standards of research practice but do not, at least in an obvious way, put at risk the integrity of research results or fair crediting of research contributions. Such acts are often seen as unprofessional behavior because they lessen the dignity of research as a profession or in some way undermine professional autonomy but are not specifically violations of research ethics. An example is the announcement of research findings in a press conference rather than by publishing in a peer-reviewed publication.[48] The argument for doubting that research integrity will be compromised is that although publication by press conference may confuse the public, other researchers will not regard such findings in the same light as results published in peer-reviewed journals.

There may even be practices that imperil research integrity but do not violate standards of research practice within a given field of research, because their effect has yet to be generally recognized. The earlier discussion of Robert Millikan's data selection methods illustrates how standards of research practice change.

Consider how the above definition of recklessness fits the James Urban case discussed earlier in this chapter, which I have taken as exemplifying the character of most misconduct cases. Let us assume that Urban's account of what he did is true: He did do the experiments in question; they are accurately reflected in the version of the paper that was published (and subsequently withdrawn); he did not intend to put fraudulent data into the literature; and his intention in submitting the fabricated data was to shorten the delay between obtaining research results and publishing them.

Suppose, however, that the experiments that he conducted had given him a significantly different result than what he had projected. Would he have had the moral courage to withdraw the paper once accepted and risk offending the editors of a prestigious journal like *Cell?* Suppose that he had died or become disabled and so could not complete the experiments, or an earthquake or other accident had disrupted his laboratory. In addition to bypassing certain "delays" built into the standard publication procedures (and so giving him an unfair advantage over any competing investigators), it also endangered the integrity of his research results, because it would have required unusual efforts to publish only genuine results in the many circumstances under which things could have gone counter to his expectation. In some circumstances, he would not have been able to act at all.

The applicability of the notion of recklessness to misconduct cases is further illustrated by a university's recent finding of misconduct against a researcher accused of plagiarism. (Plagiarism, the appropriation of another's ideas or writings and representation of them as one's own, will be discussed in detail in Chapter 9.) The accusation was made against an investigator in chemistry for using text from published articles, without attribution, in his grant proposal to NSF. The investigator had copied the work of others verbatim into his notes without quotation marks or attribution. As a result he could not distinguish his own work from that of others when he came to use his notes in writing a grant proposal. Although this was not deliberate misrepresentation of another's work as the perpetrator's, it was not simply a careless mistake, either. Dropping some quotation marks in transcribing some notes would have been a careless, perhaps even a negligent mistake. In this case, however, there were no quotation marks to lose. The failing was one of recklessness even if it was not the deliberate theft of ideas and words. In its finding of misconduct (but not plagiarism) against the subject, his university said that the subject had displayed "a reckless disregard for appropriate pro-cedures of scholarship" and had "knowingly and repeatedly [engaged] in a pattern of research note taking that, given enough time, was in-evitably going to produce precisely the situation that arose with his NSF grant proposals."[49]

Reckless research resembles other acts of recklessness, such as driving too fast to keep control of the vehicle. It involves taking a risk that, as an ethical matter, ought not be taken. Not only is recklessness rather than "fraud" in the strict sense at the heart of most research misconduct,[50] but "reckless research" also underlies many of the lesser abuses that the NRC panel that authored *Responsible Research* calls "questionable re-search practices," but which might better be called "objectionable re-search practices." Many of these practices are exemplified in the cases in the Association of American Medical Colleges' (AAMC) recent collec-tion. They include, along with clear acts of falsification and plagiarism, such things as taking unfair advantage of one's position as a reviewer of manuscripts or grants, making one's data look better than they are although without clearly falsifying them, and giving less credit to one's sources than they deserve. The AAMC cases were developed for biomedical fields, and the abuses they catalogue seem more prevalent in the biological and medical research fields (the biomedical fields have also been the site of the majority of cases of clear misconduct[51]), but many of the cases can readily be generalized across disciplines. What is important here is the empirical richness of the AAMC collection, which was assembled to reflect actual events.

SELF-DECEPTION AND RESEARCH MISCONDUCT

The introduction discussed self-deception as the failure to spell out, even to oneself, what one is doing, in circumstances under which it would be normal to do so. This characterization is superior to some others that have been offered,[52] in that it makes it clear both why self-deception is regarded as a moral failing and why self- deception, like ordinary deception, entails a failure to tell the truth.

The moral failing of self-deception should not be confused with the psychological fact of observer bias. As researchers have become aware of the potential of the normal human tendency to see what one expects to see, a type of observer bias, science has introduced refinements in methodology to control for that bias. For example, in research in which the effect of some variable is being tested, where possible, the person making the observations should be ignorant of whether the observations are of the experimental individuals who were subject to the variable in question or of members of a control group. (Where human subjects are also kept ignorant of whether they are in the experimental group or the control group, the experiment is called a "double-blind" experiment.) Today, when we recognize at least some sources of observer bias, to fail to control for such bias would be self-deceptive, not because the bias itself is self-deception, but because to fail to control for such bias is.

The 1989 edition of *On Being a Scientist* confused observer bias with self-deception: it thus makes it appear that previous generations of scientists who were influenced by observer bias that we recognize, but they did not, were all self-deceived. However, where *no one* recognizes the phenomenon of observer bias, an individual investigator is not self-deceived for failing to control for it.

A clear example of self-deception in the recent history of research is the case in which someone on the Pons and Fleischmann team changed the observed value of a signal line on the gamma-ray spectrum to one the investigator took to be in better accord with the neutron production he interpreted it to be.

On March 23, 1989, Stanley Pons and Martin Fleischmann announced a remarkable breakthrough: They had accomplished cold nuclear fusion in their University of Utah laboratory. Pons and Fleischmann's evidence was based on a graph that recorded gamma-ray emission when a cube of palladium had "melted and partly vaporized" (Taubes, 1993, p. 4). An electric current had passed through a palladium electrode and a platinum electrode in a beaker of heavy water and lithium. The gamma-ray energy showed a peak at 2.22 MeV, the unique amount of energy released when a neutron is captured by a proton, which can only happen if a neutron is generated. Despite (or due to) these straightforward results and apparent proof of cold fusion, certain scientists were not convinced.

In fact, the research and results that Pons and Fleischmann and their supporters found proved to have many discrepancies. One critic was Richard Petrasso. Petrasso's group requested that Fleischmann et al. show the full gamma-ray spectrum that were observed. It revealed that the signal line had an energy of 2.496 MeV, not 2.22 MeV. In addition, the line lacked a "Compton edge," a specific pattern that occurs due to the way gamma rays react to the detectors, which should have been evident at 1.99 MeV. Petrasso pointed out that the gamma-ray line identified by Fleischmann et al. was two times smaller than the resolution of the measuring instrument would permit. Other scientists showed that the pressure claimed by Fleischmann et al. was a miscalculation. The actual pressure was too low for fusion to occur.[53]

Had cold fusion as originally envisioned proven to be possible,[54] it would have meant that there was a relatively inexpensive route to releasing energy from fusion. This would have had enormous economic (and military) implications. In the case of research with implications as momentous as cold fusion, no one would expect to engage in falsification and have the act go undetected, because the results would be scrutinized in excruciating detail. The only explanation of the reckless act of changing that value is to assume that the investigator or the team was so fixed on believing that they had observed cold fusion that they felt sure that they knew what the data should be and deceived themselves about what they were doing in changing the value of the signal line.

Finally, let us return to the case of Millikan, since it is important to recognize and learn from the problems that have occurred in even the strongest research. I said earlier that allowing intuition to have a role in science leaves room for self-deception. Now that we have reviewed the nature of self-deception, we have the tools to consider what steps can be taken to limit self-deception without crippling intuition.

SCENARIO: INTUITION AND DATA SELECTION

You are a mature investigator in a position similar to that of Millikan in 1912; that is, you recognize that something strange sometimes occurs in your experiment. You have some ideas about the factors that are interfering and those ideas have enabled you to improve your experimental setup, but you have not yet fully identified and eliminated the causes of the episodes of strange behavior. On those occasions when you recognize that the equipment is behaving strangely, the data that you obtain for that experimental run turn out to be quite different from the data you obtain when the equipment has not been behaving strangely. You are aware that your developing understanding of the underlying phenomena, an understanding that has had some independent confirmation, developed only because you were willing to reason about the phenomena from the data that you had obtained from the experiments that did not look aberrant to you. Therefore, you have confidence in your intuition that the data obtained from the aberrant-looking experiments are not as important. You want to be honest about your method, but you are finding it difficult to describe that method.

How should you write up your experiments?

This scenario poses a problem that is not insoluble. The solutions are probably not simple, however, and they do not guarantee success either in identifying the underlying phenomena or in putting one's work before one's peers.

CONCLUSION

In summary, the notion of "scientific fraud" in the strict sense of intentionally attempting to deceive others into believing something about nature that the perpetrator does not believe to be true is not applicable to most research misconduct. Researchers often use the term "cutting corners" to refer to departures from good research practice that endanger research integrity. "Cutting corners" does not convey what is morally objectionable about those abuses, however, some of which reach the level of research misconduct. What is objectionable about them is that they are a negligent or reckless dereliction of research responsibility.

This conclusion has important practical implications. Findings of misconduct have often been stymied or overturned by the difficulty of proving intent. Such proof would not be necessary if it were recognized that certain recklessness, and not necessarily intent to deceive, is sufficient for misconduct.[55] Furthermore, recognition of recklessness in research helps identify the causes of research misconduct and, therefore, the means of lessening its incidence. As Efraim Racker points out, unusual pressure is rarely a feature of the few documented cases of fraud in the natural sciences.[56] Pressure to prove the value of one's work by publishing is a common element in the many cases of reckless research.[57] Attention to care and concern, or on the negative side, to recklessness and negligence, as well as to truthfulness and intentional deceit, is required for an adequate understanding of responsible and irresponsible research behavior.

Research ethics is less mature as a topic of discussion than is engineering ethics. Until recently, concern with extreme (and rare) behaviors have predominated. Attention to those extreme cases was important because they were frequently mishandled, and this mishandling showed the inadequacy of the research community's means of controlling wrongdoing. In particular, those who had reported such misdeeds had often suffered for doing so. However, the understanding of what is required for trustworthy behavior in research is much more complex than simply preventing the most egregious betrayals of that trust. Research practice poses many ethically significant problems for the investigator. Like design problems, these problems require complex judgment.

NOTES

1. Based on a scenario by Arun Patel and Ravi Patil, MIT '92.
2. Based on a scenario contributed by an MIT graduate student with new technical particulars added by Albert R. Meyer, EECS, MIT.
3. For a cautionary tale based on a *graduate student's* fabrication of information on a grant proposal and subsequent dismissal from his graduate program, see "Fabrication in a Grant Application" in the section on misconduct in the latest edition of *On Being a Scientist*, 1995. Second edition. Washington, DC: National Academy Press. This story may be found on the www version of *On Being a Scientist*. at http://www.nap.edu/nap/online/obas/contents/misconduct.html #Fabrication.
4. Professional responsibility is generally discussed affirmatively, rather than with the principal ways one might be derelict in fulfilling that responsibility. That

society has always been concerned about catastrophic accidents has encouraged engineers to consider their responsibility for safety may partly explain the difference.

5. Charles Perrow in *Normal Accidents* (New York: Basic Books, 1984) argues that in highly complex systems, some catastrophic accidents are "normal."

6. Petroski, Henry. 1985. *To Engineer is Human.* New York: St. Martin's Press.

7. See the papers and discussion in the "Peer Review" section of *Ethics and Policy in Scientific Publication*, 1990, by the Committee on Editorial Policy (John C. Bailar, Marcia Angell, Sharon Boots, Karl Heumann, Melanie Miller, Evelyn Myers, Nancy Palmer, Sidney Weinhouse, and Patricia Woolf) of the Council of Biology Editors (CBE). Bethesda, MD: Council of Biology Editors, Inc. pp. 257–284.

8. For brief summaries of misconduct cases that occurred at universities and that included both bizarre and typical cases, see Mazur, Allan. "The Experience of Universities in Handling Allegations of Fraud or Misconduct in Research." *Project on Scientific Fraud and Misconduct: Report on Workshop Number Two.* Rosemary Chalk. AAAS, 1989, pp. 67–94.

9. This scenario is an engineering variant of "The Selection of Data" in the second (1995) edition of *On Being a Scientist.*^{www}

   A case of a graduate student who omitted mention of the recalcitrant data is presented for ethical evaluation in NSPE Case 85-5.^{www} It was originally published in *Opinions of the Board of Ethical Review*, Volume VI, Alexandria VA: National Society of Professional Engineers, 1989, pp. 27–29.

10. Babbage, C. 1830. See discussion of Babbage in Jackson, C. Ian, 1992, *Honor in Science*, Research Triangle Park, NC: Sigma Xi, the Scientific Research Society.

11. Norman, Colin. 1988. "Stanford Inquiry Casts Doubt on 11 Papers." *Science*, 242(4 November 1988):659–661.

12. See Note 1 above.

13. Donald Buzzelli of the Office of the Inspector General at NSF argues that "falsification, fabrication and plagiarism" were intended as *examples* of "serious deviations from accepted practice" rather than defining instances of such deviation, a point that the "other serious deviation" clause simply spells out. See comments by Donald Buzzelli on the definitions of research misconduct in the Research Ethics section of the WWW Ethics Center for Engineering and Science, ethics.cwru.edu/research.html.

14. Idem.

15. Committee on Academic Responsibility Appointed by the President and Provost of MIT, *Fostering Academic Integrity.* Massachusetts Institute of Technology, 1992.

16. National Research Council Panel on Scientific Responsibility and the Conduct of Research. 1992. *Responsible Science: Ensuring the Integrity of the Research Process*, Volume I, Washington, DC: National Academy Press, p. 5.

17. I am indebted to Donald E. Buzzelli for bringing this point to my attention.

18. Kiang, Nelson. 1995. "How are Scientific Corrections Made?" *Science and Engineering Ethics* 1(4)(October):347. Guertin, Robert. 1995. Commentary on "How are Scientific Corrections Made? (by N. Kiang)," *Science and Engineering Ethics* 1(4)(October):357. Pfeifer, Mark P. and Gwendolyn L. Snodgrass. 1990. The Continued Use of Retracted, Invalid Scientific Literature. *Journal of the American Medical Association*, 263(10):1420–23.

19. Office of the Inspector General, National Science Foundation. 1993. Semiannual Report to the Congress. Number 9: April 1,1993–September 30, 1993. p. 26.
For more on this controversy, see Donald E. Buzzelli's (1993) "The Definition
of Misconduct in Science: A View from NSF," *Science*, 259:584–648 and his comments on misconduct definitions in the Research Ethics section of the WWW
Ethics Center for Engineering and Science.
20. Office of the Inspector General, National Science Foundation. 1993. Semiannual Report to the Congress, Number 9: April 1, 1993–September 30, 1993,
p. 26.
21. They also referred the case to the U.S. Attorney General, who declined to pursue
legal action. Decisions by the U.S. Attorney General about pursuing legal action
are influenced by many factors, including the costs of bringing the case and
whether the cost is a good use of the taxpayers' money given the likelihood of
conviction.
22. Jocelyn Kaiser, "Commission Proposes New Definition of Misconduct," *Science*,
269(29 September, 1995):1811. Some objection to the commission's report is
due to the variety of issues that it seeks to address simultaneously.
23. Broad and Wade, pp. 153–157.
24. Nova. 1988. *Do Scientists Cheat?* (videotape). Northbrook, IL: Coronet Film and
Video.
25. Committee on the Conduct of Science of the National Academy of Sciences.
1989. *On Being a Scientist.* Washington, DC: National Academy Press, p. 15.
26. For a discussion of Section 493 of the Public Health Service Act and subsequent
response to it, see Semiannual Report to the Congress, Number 9 (April 1,
1993–September 30, 1993) by the Office of Inspector General, National Science
Foundation.
27. Barbara J. Culliton. 1988. "Harvard Tackles the Rush to Publication." *Science*,
242(July 29, 1988):525 and Eliot Marshall, "Fraud strikes Top Genome Lab."
Science 274(8 November 1996):908.
28. National Academy of Sciences Panel on Scientific Responsibility and the Conduct of Research. 1992. Cited in Note 16 above, p. 25.
29. Goodstein, David. 1990. "Scientific Fraud." In *Engineering & Science*, Winter
1991, pp. 11–19. An excerpted version of a similar article by Goodstein was
reprinted in *The Scientist*, March 2, 1992, pp. 11, 13. Fraud legislation, for example, laws against consumer fraud, have somewhat different evidentiary criteria.
30. L. S. Hearnshaw. *Cyril Burt, Psychologist* (London: Hodder and Stoughton, 1979)
quoted in Broad and Wade, 1982, p. 209. Hearnshaw's view has not gone
unchallenged – see, for example, A. R. Jensen, 1992. "The Cyril Burt Scandal,
Research Taboos, and the Media." *The General Psychologist*, 28(Fall):16–21. However, such purported refutations have themselves been refuted – see, for example, Samelson, Fritz. 1992. "Rescuing the Reputation of Sir Cyril [Burt]."
Journal of the History of the Behavioral Sciences, 28:221–133 and the references
therein.
31. The quantity 0.771 was given in three separate studies published in 1955, 1958,
and 1966 as the correlation of IQ scores of his fictitious twins. See Leon J.
Kamin. 1974. *The Science and Politics of I.Q.* (Potomac MD) and Lawrence Earl-
baum quoted in Broad and Wade, 1982, p. 206.

32. Roberts, Leslie. 1991. "Misconduct: Caltech's Trial by Fire." *Science*, 253:1344–1347.

33. Racker, Efraim. 1989. "A View of Misconduct in Science." *Nature*, 339(May 1989):91–93.

34. Marshall, Eliot. 1996. "Fraud Strikes Top Genome Lab," *Science*, 274(8 November):908–10.

35. Holton, G. 1978. "Subelectrons, Presuppositions, and the Millikan-Ehrenhaft Dispute," first published in *Historical Studies in the Physical Sciences*, 9:161–224 and reprinted in his book *The Scientific Imagination*, pp. 60–61. My page references to this paper of Holton's are to the reprint in *The Scientific Imagination*.

36. See Holton's "Subelectrons, Presuppositions, and the Millikan-Ehrenhaft Dispute" and his forthcoming article "On Doing One's Damnedest: The Evolution of Trust in Scientific Findings," Chapter 7 in Holton's *Einstein, History, and Other Passions*. New York: American Institute of Physics.

37. Holton, Gerald. 1994. "On Doing One's Damnedest: the Evolution of Trust in Scientific Findings," Chapter 7 in Holton's *Einstein, History, and Other Passions*. New York: American Institute of Physics.

38. Quoted by Gerald Holton. "Subelectrons, Presuppositions, and the Millikan-Ehrenhaft Dispute," p. 63, from Robert A. Millikan. 1913. "On The Elementary Electrical Charge and the Avogadro Constant," *Physical Review*, 2:109–143.

39. Allan D. Franklin. 1981. "Millikan's Published and Unpublished Data on Oil Drops," *Historical Studies in the Physical Sciences*, 11:185–201.

40. Broad, William and Nicholas Wade. 1982. *Betrayers of the Truth*. New York: Simon & Schuster.

41. Committee on the Conduct of Science, 1989, p. 14.

42. National Academy of Sciences Panel on Scientific Responsibility and the Conduct of Research. 1992. Cited in Note 16 above, p. 26.

43. See also the previously cited booklet from the Center for the Study of Ethics in the Professions Module Series by Martin Curd and Larry May, *Professional Responsibility for Harmful Actions* (Dubuque Iowa: Kendall/Hunt, 1984). These authors discuss retrospective responsibility ("judge problems") and criteria for ethical and legal fault in engineering practice.

44. Taubes, Gary. 1993. *Bad Science*. Random House: New York, p. 142.

45. Ibid., p. 142.

46. Taubes, *Ibid.*, pp. 142–143, 167–174, 253–254. The NSPE Code of Ethics has a provision directly applicable to the saga of Marvin Hawkins: "10. a. Engineers shall, whenever possible, name the person or persons who may be individually responsible for designs, inventions, writings, or other accomplishments."

47. Shore, Eleanor. 1995. "Effectiveness of Research Guidelines in Prevention of Scientific Misconduct." *Science and Engineering Ethics*, 1,4(October):383.

48. This is a practice that the NAS panel classifies as a "questionable research practice" *op. cit.*, p. 28.

49. Office of the Inspector General, National Science Foundation. 1992. Semiannual Report to the Congress No. 7, April 1, 1993–September 30, 1993, p. 37.

50. In addition to the sources already cited, the reader will find descriptions of misconduct cases frequently written up in *Science* and in journals of the disciplines in which they occurred. The Office of the Inspector General of NSF issues a

semi-annual report to the Congress that includes a section on oversight dealing with misconduct cases and other audits and inspections conducted by that office.

51. Patricia K. Woolf. 1986. "Pressure to Publish and Fraud in Science," *Annals of Internal Medicine*, 104(2)(February):254–256.

52. For example, see the first (1989) edition of *On Being a Scientist*, which treated self-deception as a manifestation of the fallibility of human perception, reasoning, and foresight.

53. Taubes, Gary. 1993. *Bad Science*. New York: Random House, pp. 3–4, 44–45, 142–143, 167–174, 253–254, and "Measurement of Gamma-Rays from Cold Fusion." 1989. *Nature*, 339(June 29):667–668. Letter and rebuttal between Fleischmann et al. and Petrasso et al.

54. Some credible experiments indicate that when the ratio of the number of deuterium (heavy hydrogen) to the number of atoms of palladium exceeds 0.85 (a ratio that is extremely difficult to achieve reliably), excess heat is consistently produced. It is still not known whether this is due to nuclear fusion. If it is, the conditions for this sort of cold fusion will have turned out to be much more difficult to achieve than Pons and Fleischman had originally suggested. (See David L. Goodstein. 1994. "Whatever Happened to Cold Fusion?" *Engineering & Science*, Fall:15–25; reprinted from the American Scholar, 63:4, Autumn 1994.) Hence the only hypothesis of cold fusion that now has any plausibility does not have the dramatic economic and military implications of Pons and Fleischman's original claims.

55. The most famous case of the overturn of a research misconduct finding because of failure to prove intent is that of the case against Mikulus Popovic and Robert Gallo whom the Office of Scientific Research Integrity of the PHS claimed had misappropriated a sample of AIDS virus sent to them by Luc Montagnier of the Pasteur Institute in France. When the case against Popovic was overturned on appeal, the Office of Scientific Research Integrity decided not to pursue its case against the head of the same research team, Robert Gallo. See Cohen, 1994, p. 23.

56. Kiang (1995) discusses an exception to Racker's generalization, a case in which a graduate student, feeling pressured by his supervisor to provide experimental confirmation of the supervisor's hypothesis, provided fraudulent confirmation.

57. Patricia K. Woolf. 1986. "Pressure to Publish and Fraud in Science," *Annals of Internal Medicine*, 104(2)(February).

7

THE RESPONSIBILITY OF
INVESTIGATORS FOR
EXPERIMENTAL SUBJECTS

A BRIEF HISTORY OF THE USE OF HUMAN SUBJECTS IN
MEDICAL AND PSYCHOLOGICAL EXPERIMENTS IN THE
UNITED STATES

The doctrine of informed consent for human experimentation was promulgated in the Nuremberg code in 1946 after the discovery of brutal experiments carried out by the Nazis.[1] It was refined in the Helsinki declarations issued by the World Medical Association in 1962 with subsequent revisions in 1964, 1975, and 1989.

The requirement of informed consent now applies to behavioral research as well as medical research. Many of the classic experiments in psychology involve deception and even clear harm to the subject and would not be allowed today.[2]

The informed consent standard was only gradually adopted in the United States, however, and it took some shocking cases in this country to demonstrate the need for reform. In 1966, a well-respected clinician and investigator, Henry Beecher, had published an article in the prestigious *New England Journal of Medicine*, in which he reported common but unethical treatment of human research subjects in many premier institutions in the United States.[3] Subsequently, the NIH developed the first Public Health Service Policy on the Protection of Human Subjects. At first these had only limited application. Later they were expanded to apply to all human subjects research conducted or supported by what was then the Department of Health, Education and Welfare.[4]

One of the most infamous experiments was the Tuskegee Syphilis Study, a study the Public Health Service had conducted from 1932 to 1972.[5] This study had started innocently enough as a treatment program for syphilis, although in the 1930s, before the discovery of penicillin, treatments were largely ineffective. When it became clear that

227

there were not enough resources to treat all those who had syphilis, the project became a study of untreated syphilis. Later, even after the discovery of an effective treatment, penicillin, the shocking decision was made to continue the study and even prevent subjects from getting treatment when they came to it by other means. The group from whom effective treatment was knowingly withheld were African-American men, sharecroppers. The medical community went for decades without raising questions about articles with titles like "The Course of Untreated Syphilis in Negro Men." Medical journal readers are unlikely to have withheld comment if the titles had been something like "The Course of Untreated Syphilis in College Students." The tolerance of the medical community for the continuance of this research evidences racism and a willingness to exploit the powerless. Since Tuskegee, the institution that conducted the study, is a historically black institution, it is not only whites who were at fault.

Another set of experiments dating from World War II that has only recently come to light are a series of radiation experiments funded by the Department of Energy (DoE). The motivation for some of these studies was to learn more about radiation injuries for the sake of workers who had been exposed to radiation in weapons work during the war. Other DoE-funded radiation studies simply used radioactive tracers, for example, to conduct nutrition studies. Some experiments met today's standards for the treatment of human subjects. In others the patients were unharmed, but their informed consent was never obtained. Still others were extremely damaging to subjects, including tests that irradiated the testicles of prisoners or subjected patients supposedly dying of cancer to massive doses of irradiation. (Some of the irradiated patients turned out not to have fatal cancer).[6]

The radiation experiments that seriously harmed patients violated not only the informed consent standard but the standard that predated it as well. As we saw in the introduction, before the adoption of the informed consent standard, the informal rule for ethical experimentation was for investigators to first do to themselves anything to which they proposed to subject others – an inverted golden rule. The radiation experiments were typical of experiments in earlier decades in using patients and prisoners as the subjects in many of the experiments that were unethical even by that earlier standard. Today regulations on human experimentation are intended both to implement the informed consent standard and to provide special protections for vulnerable populations who have been the most subject to abuses in the past. The DoE radiation tests have prompted numerous lawsuits against hospitals and universities.[7]

REQUIREMENTS FOR THE USE OF HUMAN SUBJECTS
IN RESEARCH

Today the ethical right to refuse to participate in experiments is coming to be regarded as an absolute right, like the right to refuse medical treatment.

U.S. federal regulations require that subjects receive a description or statement of

1. the purpose of the experiment and the procedures to be used,
2. foreseeable risks and discomforts,
3. foreseeable benefits,
4. appropriate alternatives, if any,
5. the extent of confidentiality of the test results,
6. the availability of medical treatment for any injury received in the experiment, and compensation for any disability,
7. whom to contact in case of questions, and
8. assurance that participation is voluntary and that neither refusal to participate nor later withdrawal from the study will result in loss of benefits to which the person is otherwise entitled (such as future care at the same facility).

The regulations also require that patients be told of any additional risks to the subject or fetus, if the subject becomes pregnant; circumstances in which a subject's participation may be terminated without their consent; any additional costs to the patient that may result from participation; the consequences of a subject's decision to withdraw and process for doing so. They are also to receive any new findings from the research that might bear on their continued willingness to participate and be informed of the approximate number of subjects in the study. Finally, the regulations forbid requiring subjects to waive any of their legal rights or to release investigators from potential wrongdoing.[8]

Institutional Review Boards (IRBs) provide key support for the exercise of that right. In the United States any institution receiving government funds must have an IRB that reviews all research protocols for experiments involving human subjects. (If investigators are applying for government funding for a particular study, that study usually must be approved by the IRB before the research can be funded.) The membership of such boards is fairly constant, and their accumulated experience prepares them to recognize dangers that an individual experimenter might overlook.

In experimentation on children or others not fully competent to consent for themselves, IRBs are alert to the danger that those who give consent as their "proxy" may have motives that conflict with protecting the interests of their wards. This might occur if guardians received payment for the participation of the subjects, for example. In many cases the consent of both the subject and the subject's guardian are required to continue the experiment. The guardian is able to give fully informed legal consent, but experimenters may also be required to stop the experiment if children or demented adults say they do not want to continue.

Because experience has shown that institutionalized subjects, such as patients or prisoners, are liable to be influenced by what they think authorities want, special measures are taken to avoid subtle coercion in recruiting them. Therefore, investigators must assure subjects that refusal to participate or withdrawal from the study will not result in loss of benefits to which the person is otherwise entitled. Other protections and assurances are specific to the situation and vulnerabilities of the subjects. Students in psychology courses used to be regularly required to be subjects in the (mostly harmless) psychology experiments of their professors. This situation continued into the 1980s. That practice is now recognized to be coercive, and students may choose alternatives to test participation.

Informed consent by research subjects is different from informed consent for medical or other treatment. The requirement of something like informed consent for treatment is much older than that for experiment; treatment without consent has long been understood in legal terms as a "battery." In both cases the mechanism of informed consent involves signing a form, supplemented by oral explanation. In the case of treatment, however, this is meant to be part of a larger enterprise of shared decision making between patient and provider. So, for example, a patient may change a consent form before signing it, say to specify that only the named surgeon, and not the surgeon's students or "associates" may perform the surgery. In contrast, research subjects cannot modify a research protocol.

Human subjects are frequently used in biomedical research. In the United States the approval of biomedical devices requires a lengthy (and costly[9]) review process by the Food and Drug Administration (FDA). Such a process is required for drugs and devices, but not for all experimental treatments. For example, the artificial heart, but not transplantation of a baboon heart, required FDA approval. Recently, the FDA halted development of a new device for emergency resuscitation of heart attack patients because testing of the device requires informed consent.[10] This consent could not be obtained from patients actually having heart attacks.

HUMAN SUBJECTS IN PRODUCT TESTING

Testing of most products is not subject to the regulations regarding the use of human subjects, since manufacturers do not receive government funds for any of their testing. An extreme example of injury in product testing occurred in October 1991 when an aircraft manufacturer, McDonnell Douglas, was reported to have caused serious injury to senior citizens recruited to test an evacuation procedure. According to *The Wall Street Journal*, McDonnell Douglas was seeking to demonstrate that it could safely increase the seating of an all-economy-class aircraft from 293 seats to 400. In response to criticisms that using employees and family members of employees as test subjects biased the test results, the company recruited senior citizens as well. Nothing resembling informed consent was obtained from the subjects. Failure to obtain informed consent was not itself prohibited by law or regulation, although the action certainly appears reckless to anyone familiar with ethical standards for human experimentation.

Even absent any familiarity with those standards, the company violated ordinary standards of prudence and responsibility when, after eleven people had been injured in the morning evacuation drill, they reportedly proceeded to conduct an afternoon evacuation drill with no warning to participants of possible injury. In all, forty-four people were reportedly injured, including eleven who were taken to the hospital, six of whom had broken bones. One sixty-year-old woman was paralyzed from the neck down due to a fracture of her spine.[11] The injured parties are certainly able to sue for damages, of course, but the point of informed consent requirements is to prevent such occurrences.

WHY EXPERIMENT ON ANIMALS TO BENEFIT HUMANS?

An alternative to some uses of human subjects is the use of nonhuman animals as subjects. Another is to forego any research that causes or makes likely serious harm to living beings. Animal studies became common after Claude Bernard, the French physiologist, founded physiology as an experimental science. He did his experiments at home, and his wife, whose ample dowry funded his work, became active in the cause of "antivivisection," as objection to such experiments was then called. Thus controversy is certainly not new, but groups such as People for the Ethical Treatment of Animals (PETA) have experienced a marked growth in numbers since the 1970s.

USING ANIMALS IN MEDICAL EXPERIMENTS

Suppose that certain experiments are thought to be needed to develop better means of coping with extreme pain in humans. These experiments would involve performing a variety of procedures that would be very painful to the experimental subjects. The subjects would have to be vertebrates. Species closer to humans in evolutionary terms might give more meaningful results, but it is not known exactly what traits of laboratory animals are most relevant. Because the information about neural response would be crucial, subjects would receive no pain medication. The subjects would be temporarily paralyzed to keep them from flailing about.

What, if any, species would it be ethically acceptable to use in such experiments? Is it morally relevant whether the subjects are mammals?

Would intelligence of members of that species be morally relevant to the decision, and if so, how?

Would the presence or absence of a complex social system in which members care for other members of the species be a morally relevant factor to consider?

Would it be morally relevant that one candidate species resembled humans more closely?

Would it be relevant that some particular individuals had once been human pets? If so, would it be better or worse to use those individuals?

A comparable experiment to the one above was a study of head trauma to restrained baboons conducted at the Experimental Head Injury Laboratory at the University of Pennsylvania. The Society for Neuroscience estimates that there are about two million cases of head trauma each year causing 50,000 deaths. The Pennsylvania study became well known when members of the Animal Liberation Front, a clandestine international group, broke into the Experimental Head Injury

Laboratory and stole videotape records of the experiments. They gave the videotape to PETA who then circulated a twenty-five-minute selection of excerpts. [12]

The question of whether or under what circumstances it is permissible to experiment on animals for human benefit is a question of the moral standing of the experimental subjects, the harm or discomfort that the experiments will cause them, and the benefit to be derived to humans. (The treatment of farm animals raises some similar issues, but that does not arise directly in the work of engineering and science, as does the use of animals in experiments.) As we saw in the introduction, for a being to have moral standing means that such an individual's well-being must be considered for its own sake. Not all beings that have moral standing need have the *same* moral standing. The existence of laws against cruelty to animals evidences the widespread conviction that animals (especially vertebrates) have moral standing, but if they were thought to have the same moral standing as humans, then having work animals would be regarded as slavery and it is not. That certain beings have moral standing does not require that they have the same moral standing as people, nor that they be said to have rights, but only that their welfare must be considered for its own sake.

The NIH requirement for Institutional Animal Care and Use Committees shows that government regulation as well as popular sentiment agrees that experimental animals do have moral standing, and vertebrates are regarded as having an especially high moral standing. The disputed question is how high? The answer to this question will bear on the determination of whether certain negative consequences achieve benefits to others. If certain animals have the same moral standing as people, then, since animals are incapable of giving informed consent, experimentation looks like unjustified risk shifting. Animal rights theorist Tom Regan believes that they do have moral standing and therefore holds that the use of animals to benefit humans is unjustifiable. [13] In contrast, Peter Singer, one of the best known theorists arguing that animals deserve better treatment, holds that *some* use of animals in research is justified when the benefits to humans are sufficiently great and not obtainable by other means.

If a being is itself a moral agent, that is commonly taken to be grounds for saying that the being has very high moral standing, since it is one criteria of being a person. Some other criteria commonly given as a basis for determining moral standing are rationality or at least intelligence, sentience (the ability to experience pain and pleasure), and being alive.

As we saw earlier, there has been a tendency for people to accord higher moral standing to those who are like themselves. Unless the similarities are morally relevant ones, this tendency is merely a self-serving prejudice.

RESPONSIBILITY FOR EXPERIMENTAL ANIMALS

Many of the particular responsibilities for the welfare of experimental animals depend upon more than the moral standing of the species or individual animal. Notice that what is permissible treatment of wild mice is very different from the permissible treatment of experimental animals. Why are the mice in the walls (whether they are wild strains or feral laboratory strains) treated so differently from the mice in the cages?

We saw earlier that responsibility stems not only from knowledge, but from a position of relationship and control. So parents have responsibilities for their children regardless of whether they know much about rearing children. The difference in the treatment of the mice in the walls and the mice in the cages is another instance. Experimental animals are under the experimenters' care and control. This position of control carries with it special responsibilities that do not apply in the case of others of the same species. In a similar way, a pet owner might be liable under legislation against cruelty to animals for doing to pets something that the pet owner would be allowed to do to animal pests.[14]

How far does an investigator's moral responsibility for animal welfare extend? Is it enough to comply with regulations, or ought an investigator to understand the habits and temperament of the species of experimental animals in his care and treat each species in accord with such things as the animals need for or aversion to touch, or their preference for either fresh or familiar bedding material? Ought an investigator take care to design experiments so as to obtain the required data from as few animals as possible? Investigators who experiment with animals have recommended that their colleagues improve animal care in these respects, even in advance of regulation.

CONCLUSION

The requirements on the treatment of experimental subjects reflect both the moral standing of people and animals and the special responsibilities an investigator takes on by establishing a relationship of some control with respect to these subjects.

NOTES

1. Caplan, Arthur L. 1992. *When Medicine Went Mad: Bioethics and the Holocaust.* Totowa NJ: Humana Press.
2. Celia Fisher has reviewed many such studies and has prepared educational materials for discussing the ethical issues they raised. For sample discussion of those classic studies see the Research Ethics section of the WWW Ethics Center for Engineering and Science and the reference there to the publication of her materials.
3. Beecher, Henry K. 1966. "Ethics and Clinical Research." *New England Journal of Medicine.*, 274(24):1354–1360.
4. Levine, Robert J. 1986. *Ethics and the Regulation of Clinical Research.* Second ed. Baltimore: Urban & Schwarzenberg.
5. Brandt, Allan M. 1978.
6. Mann, Charles C. 1994. "Radiation: Balancing the Record." *Science,* 263(5146): 470–474.
7. Wheeler, David L. "Making Amends to Radiation Victims." *The Chronicle of Higher Education,* 10/13/95.
8. Levine, Robert J. 1995. "Informed Consent: Consent in Human Research." *Encyclopedia of Bioethics,* 2nd ed. New York: Macmillan, pp. 1241–1250. See also the online materials on requirements available through the WWW Ethics Center for Engineering and Science.
9. For example, Trimedyne, which spent $2 million to develop a device to use lasers to vaporize fatty deposits in coronary arteries, waited from 1983 to 1993 for FDA permission to use it in diseased leg arteries. Finally, they received permission only for "no risk, no benefit" use in leg arteries (in a patient who was to have artery grafts for other reasons).
10. Winslow, Ron. 1994. "FDA Halts Tests on Device That Shows Promise for Victims of Cardiac Arrest." *Wall Street Journal,* Wednesday, May 11, 1994, p. B8.
11. Nazario, Sonia L. 1991. "McDonnell Douglas Jet Evacuation Drills Leave 44 Injured." *Wall Street Journal,* Wednesday, October 30, p. A3–A4.
12. Tannenbaum, Jerrold and Andrew Rowan. 1985. "Rethinking the Morality of Animal Research." *Hastings Center Report,* October, pp. 32–36.
13. Singer, Peter. 1985. *In Defense of Animals.* New York: Blackwell; Singer, Peter. 1977. *Animal Liberation.* New York: Avon Books; Singer, Peter. 1990. "The Significance of Animal Suffering." *Behavioral and Brain Sciences,* 13(1):9–12; and Regan, Tom. 1983. *The Case for Animal Rights.* Berkeley, CA: University of California Press.
14. The extermination of pests finds justification in the threat they pose for spreading disease and destroying property. The moral standing of some species, especially vertebrates, is often argued to require using methods that force wild pests to relocate elsewhere or at least kill them with a minimum of suffering.

8

RESPONSIBILITY FOR
THE ENVIRONMENT

SUSPECTED HAZARDOUS WASTE [1www]

You are an engineering student employed for the summer by a consulting environmental engineering firm. R. J., the engineer who supervises you, directs you to sample the contents of drums located on the property of a client. The look and smell of the drums and samples leads you to believe that analysis of the sample will show it to be hazardous waste. You know that if the material contains hazardous waste, there are legal restrictions on the transport and disposal of the drums that will apply and that federal and state authorities must be notified.

When you inform R. J. of the likely contents of the drums, R. J. proposes to "do the client a favor" – document only that samples have been taken and not proceed with the analysis. R. J. further proposes to tell the client where the drums are located, that they contain "questionable material," and to suggest that they be removed.

Why does R. J. think that incomplete information will be "a favor" to the client?

Is giving incomplete information responsible engineering practice?

Does the law in your state require engineering firms to report any release of hazardous waste to the state's department of environmental protection?

What, if anything, could you do, as a student and a summer hire?

COMPLYING WITH POORLY WRITTEN
ENVIRONMENTAL LAWS

You are an environmental engineer working for a manufacturing company that makes computer components. In the process your plant

236

creates toxic wastes, primarily as heavy metals. Part of your job is to oversee the testing of the effluent from your plant, signing the test results to attest to their accuracy, and supplying them to the city.

The allowable levels of heavy metals in the effluent are intended to be several times as stringent as federal law allows, because of recreational use, including swimming and fishing, downstream. However, the law was poorly written. It limits the *concentration* of toxic material in the effluent rather than the *quantity* discharged in a given period. Therefore, the requirement can always be met by diluting the discharge.

Although you are complying with the law, you are concerned that the increased amount of heavy metals you have begun putting in the river may pose a health hazard, especially to some of the residents who regularly eat catfish caught downstream.

What can and should you do, and how should you go about it?[2]

THE RISE OF ECOLOGY

Engineers' and applied scientists' responsibility for the environment in some respects resembles the responsibility for safety, but new ways of thinking about the environment show the matter to be quite complex. New thinking about the environment has emerged rapidly since the early 1960s.

A few decades ago "the environment" meant simply the surroundings, the assemblage of stuff nearby. The idea of the environment as an integrated system has only been in wide use since the 1960s, although some argue that it resembles notions of nature or the Earth found in many cultures originating outside Western Europe, or notions of nature that predate modern science.[3] The present view of the environment arose with a new scientific discipline, ecology: the study of the relationship between organisms and their environment.

In English, the terms "environment," "ecology," and "ecosystem" have come into common usage to convey the ideas of integrated systems in nature. "Ecology" names a field of study sufficiently obscure that in 1971 the term was not even included in that year's edition of the authoritative *Oxford English Dictionary*. The emergence of the science of ecology has given rise to many new areas of engineering theory and practice (especially related to chemical engineering and civil engineering) that are called "environmental engineering."

The conservation movement had existed for many decades. It had been popularized by Theodore Roosevelt in his term as president. Conservation efforts bore some resemblance to today's efforts to protect or improve the environment, but they were directed toward the preservation of specific entities: recreational areas or natural resources of evident economic significance, such as forests, fish stocks, and navigable waterways. Conservation efforts proceeded without a comprehensive understanding of the relationships between organisms and their environments.

RACHEL CARSON[www]

Rachel Carson, a marine biologist, was the person who did most to change thinking about what we now call "the environment." Besides being a meticulous scientist, she had won recognition as a science writer even before her famous book on the effects of pesticides, *Silent Spring*. Carson had published *Under Sea Wind* in 1941, and in 1951 she won the National Book Award for *The Sea around Us*. Concerned by the growing evidence of major damage to fish, birds, and other animal life as a result of new mass applications of pesticides to large areas of wilderness, Carson sought someone to write a book about the subject. This was a task that she finally took on herself in *Silent Spring*. This work, published in 1962, changed the consciousness both of the public and of policy makers about the effect of pesticides on the environment.[4] As John George said regarding the effects of this book, it

> had sufficient impact to reach the highest levels of government, and these decision-makers began to listen to the idea that ecological process was vital to life and to our own well-being. As a result, the ecologist was less easily dismissed as well meaning but completely unrealistic.[5]

The characteristics demonstrated by Rachel Carson in her efforts to bring the danger of pesticides to public attention are similar to those demonstrated by Roger Boisjoly and by William LeMessurier in their attempts to bring attention to the safety hazards that they recognized. Among these virtues are the courage to tell the truth that no one wants to hear, even if it means being harshly criticized; concern for fairness to everyone; and a concern for the safety and well-being of others.

The character of Carson's goal was rather different from that of Boisjoly and even LeMessurier, as were her circumstances. Boisjoly's purpose was to alert the decision makers in his company or at NASA to the danger of a fatal explosion. LeMessurier sought to find a way of remedying a flaw in structural supports for the huge Citicorp Tower. Repairing the Citicorp Tower and safeguarding the public in the meanwhile required enlisting the aid of *many* individuals and organizations. Carson's task was to reverse a major trend in social policy regarding the use of chemical pesticides. It would not have been sufficient for Carson to have her employer in the 1940s, the U.S. Fish and Wildlife Service, appreciate the problem. Many people in that service had in fact already become aware of the unexpectedly severe harm to wildlife caused by new pesticide use.

To accomplish the reversal, she needed to counter the enthusiasm for pesticides that had grown out of their use in World War II to combat insect-borne diseases. DDT in particular had proved very effective in controlling lice that had formerly spread disease such as typhus among troops in wartime, killing more members of the armed forces than did wounds.[6] The great enthusiasm for insecticides in the postwar period was reflected by the award of the Nobel Prize for Physiology to Paul Hermann Müller of Switzerland in 1948 for his 1942 discovery of the insecticidal properties of dichlorodiphenyltrichloroethane, later known as DDT. (DDT had first been synthesized in 1874.)

After World War II, pesticides and the goal of "eradicating" – as contrasted with simply controlling – insect pests gained wide popularity. By the late 1950s this led to massive use of chemical pesticides, including government spraying of vast tracts of land without obtaining the consent of residents.

Carson realized the problem required first making it generally understandable to both the voting public and governmental policy makers. Accomplishing this goal required the integration of a diffuse and variable body of data into a clear and compelling account.[7] It also required that she not be dependent on keeping her job, since, if effective, her book was sure to raise a storm of protest. It is interesting to compare Roger Boisjoly's efforts to get his company to appreciate the hazard posed by the performance of the *Challenger's* O-rings in cold weather and William LeMessurier's initiative to mobilize private and public officials to protect the public safety while the structural supports of the Citicorp Tower were strengthened, with Carson's campaign to change public consciousness and the differences their goals made to the way each went about his or her task.

KEY ENVIRONMENTAL LEGISLATION

Before the late sixties there was little significant national or state legislation to protect the environment.[8] Individuals could bring lawsuits, but rarely was a single party so harmed by pollution as to take this route.

Below is a listing of major U.S. environmental legislation through 1986. Each measure had its own history, which we cannot consider here, but taken together, these acts represent an extraordinary shift from the 1950s.

1969: National Environmental Policy Act requires environmental impact statements for actions by federal agencies that affect the environment. Congress then created the Environmental Protection Agency (EPA) to enforce compliance with this law.

1970: Occupational Safety and Health Act established the National Institute of Occupational Safety and Health (NIOSH) within the Department of Health and Human Services (DHHS). NIOSH develops mandatory health and safety standards for business, conducts research on occupational health problems, and produces criteria identifying toxic substances and safe exposure levels for them.

1970: Clean Air Act, amended in 1977 and 1990 with National Emission Standards for Hazardous Air Pollutants (NESHAPS), regulates off-site contamination, that is, pollution outside one's facility.

1972: Clean Water Act, amended in 1972, 1977, 1986, and 1995, applies to off-site contamination.

1976: Resource Conservation and Recovery Act (RCRA) provides regulation for the on-site handling of toxic chemicals, that is, handling of toxic chemicals at one's facility.

1976: Toxic Substances Control Act (TOSCA) provides regulation to protect the public against toxic substances in consumer and industrial products.

1980: Comprehensive Emergency Response, Compensation and Liability Act (CERCLA), popularly known as "the Superfund Act," established the Agency for Toxic Substances and Disease Registry (ATSDR) within the Public Health Service (PHS). The PHS is an agency of the DHHS.

1986: Superfund Amendments and Reauthorization Act (SARA) added to the duties of the ATSDR the responsibility for developing and updating a list of toxic substances that pose the most significant threat to human health.[9]

THE CONCEPT OF AN ECOSYSTEM

The word "ecosystem" has entered the general English vocabulary from the technical vocabulary of ecology. An ecosystem comprises a group of organisms that interact with each other and with their physical environment in ways that affect the population of those organisms.

Thinking in terms of ecosystems (or in terms of systems generally) directs attention away from particular individuals or species to the way its interactions sustain and are sustained by the whole system, even if one's interest is primarily in that individual or species.

An example of a simple ecosystem is that of kelp, sea urchins, and sea otters. Kelp is a commercially valuable ocean plant used in foods, paints, and cosmetics. The kelp forests along the Pacific coasts grow in long streamers attached to the ocean floor. In recent years the kelp suddenly began disappearing. Concurrently, sea urchins had increased because the population of sea otters, a major predator of sea urchins, had fallen off drastically. The sea urchins feeding on the kelp weakened its attachment to the ocean floor, causing it to float away. Understanding the kelp forest as one part of an ecosystem showed that the simplest way to protect the kelp forests was to protect the sea otter. When the population of sea otters flourished, the sea otter held down the population of sea urchins, and the kelp forests were restored.[10]

Ecosystems vary in their fragility. An ecosystem is more resilient if several component species perform the same function or "fill the same niche" in the system. The kelp–sea urchin–sea otter system would have been more resilient if sea urchins had had more natural predators. Then when the otter population diminished, the other predators would have fed on the more plentiful sea urchins, perhaps before they threatened destruction of the kelp.

The systems approach to understanding biological phenomena transforms the ways in which we think of those phenomena. Systems may be naturally occurring, like an ecosystem or the digestive system of an organism; or they can be the product of human endeavors, like transportation systems, political systems, or educational systems. Systems comprise sets of components that work together to perform a function, yet the system is more than the sum of its components. The continued function of the whole system is the central concern, rather than the flourishing of any single component or components.

As we saw in the introduction, for a being to have moral standing means that such an individual's well-being must be considered for its own sake. Ecological thinking considers the good of one individual or species only in relation to the whole ecosystem. Therefore, concern for the environment does not come down to consideration of the moral standing of individuals or of species. Systems thinking complicates consideration of harms and benefits and the risk or probability of their occurrence.

HAZARDS AND RISKS

Risk is commonly used as a term for a danger that arises unpredictably, such as being struck by a car. Sometimes it is used for the *likelihood* of a particular danger or hazard, as when someone says, "You can reduce your risk of being hit by a car by crossing at the crosswalk."

In the context of "risk assessment" or "risk management," "risk" has a technical sense: the probability or likelihood of some resulting harm multiplied by the magnitude of the harm. One can then compare, say, the risk of death for a driver under 5' 4" tall from a deploying airbag with the risk of death from being in an automobile accident unprotected by airbags. (Shorter drivers are at greater risk of injury from airbags, because they sit closer to the wheel.) One could also compare the risk of harms or the likelihood of benefits of different magnitudes. In the introduction we considered two monetary risks: the rather common event of losing change in a broken vending machine and the comparatively unlikely loss of one's wallet due to robbery at gunpoint. The use of the technical sense of risk requires that one be able to meaningfully quantify the resulting harms.

Financial loss, risk of death, and days of illness are commonly used quantifiable harms in risk analysis. Many other harms are difficult to quantify except in an arbitrary way, however. Therefore, cost-benefit and risk-benefit analyses should be understood as techniques that help weigh a restricted class of consequences. They are not general techniques for determining what course of action produces the greatest good in a situation, as they are sometimes claimed to be. They may help agents devise a responsible course of action for those hazards that are tolerated to some degree. The tolerance of some environmental risk is implicit in the authorization of the Environmental Protection Agency (EPA) to regulate (only those) chemicals that pose an *unreasonable* risk of injury to health or to the environment. As we saw in Chapter 3, however, there is zero tolerance for food additives that pose *any* known risk to humans or are known to be carcinogenic to other species.

When making formal risk calculations, it is important to be careful not to ignore relevant considerations that cannot be represented as arithmetic quantities. An example we considered earlier is the emotional trauma associated with the loss of money at gunpoint, a factor we did not consider in comparing the risk of monetary loss to a vending machine and to a robber. Our cost-benefit calculation did not tell the whole story, but because of the widespread use of such techniques, it has become important to quantify environmental effect where possible.

The professional responsibility of engineers and applied scientists for environmental protection, like their responsibility for ensuring public safety, requires attention to two sorts of risks:

- hazards that have gone unrecognized, at least by some key decision makers, and that pose a grave or excessive threat to safety or the environment; and
- hazards that are a recognized feature of the situation but cannot be completely eliminated and are mitigated only by increasing other risks and costs.

Engineers will vary in their ability to recognize previously unrecognized hazards (hazards of the first type) and to mitigate known hazards (hazards of the second type); but addressing the two types of hazards raises different issues.

Environmental hazards that have gone unrecognized are analogous to the safety hazards posed by cold temperatures to the performance of the O-ring seals in the *Challenger* booster rockets, increasing the load on the rods supporting the walkways of the Kansas City Hyatt Regency, and the unrecognized toxicity of some new chemical. The threat of accident is a typical hazard of this sort, but, as the example of toxicity of a new chemical illustrates, it is posed not only by accidents. What is essential is that the threat is unrecognized or disregarded. For these the ethical problem is to bring such risks to light and have them addressed.

Engineers and applied scientists, because of their education and training, are in a special position to recognize both environmental hazards and safety hazards. This is the basis for the growing consensus that engineers and applied scientists have a professional responsibility to bring environmental as well as safety hazards to light. So it is that engineers – not only environmental engineers or those who have environmental protection as an assigned (i.e., official) responsibility – are widely acknowledged as having a professional moral responsibility to prevent environmental damage. For example, the IEA states that its members should practice "in accord with sustainability and environmental principles."

The IEEE in its most recent (1990) code of ethics states as the first of the ten points to which IEEE members are committed:

> to accept responsibility in making engineering decisions consistent with the safety, health and welfare of the public, and to disclose promptly factors that might endanger the public or the environment...

It is interesting that the IEEE has been in the forefront on this question, although electrical engineers are less likely than civil engineers or chemical engineers to be environmental engineers.

Recognized and unrecognized hazards differ in the extent to which engineers may be prepared or assigned to consider such hazards. Techniques for coping with specific recognized hazards are a part of the subject matter of many engineering disciplines. A particular engineer may or may not be proficient in any particular technique. A professional responsibility not to take work beyond the limits of one's competence is widely acknowledged and is explicitly stated in the codes of ethics of several engineering societies. For example, the second of seven "fundamental canons" in the Code of Ethics of the American Society of Mechanical Engineers states:

> Engineers shall perform services only in the areas of their competence.

One can undertake the task of checking for overlooked factors that may cause an accident. We saw that structured techniques such as fault-tree analysis can be used in estimating the probability of accidents. Engineers should take on such work only if qualified to use these techniques. But it does not make sense to object that some engineer is unqualified to raise a concern about some previously undetected hazard. Of course, it is possible that the concern will, after more expert assessment, prove groundless, but the engineer would not have been presuming to provide services beyond his or her competence simply by raising the issue.

The situation is rather different with assessment of the risks from recognized hazards. Since they are recognized, there is not the same issue of warning about them. Furthermore, because they have been recognized, but persist, they are usually ones that cannot be eliminated or mitigated without incurring other significant risks and costs. For these the task is to evaluate the risk, find ways of mitigating it, and evaluate the risks and costs of actions that would mitigate the first risk. Here the issue of performing services only within limits of one's competence does become a significant factor.

Illustration from the Exxon Valdez Oil Spill Case

The contrast between the two sorts of hazards may be further illustrated by considering two of the factors that contributed to the catastrophic oil spill when the Exxon oil tanker Valdez went aground on a reef in 1989. First, a seaman had previously warned about alcohol abuse by the captain of the Valdez. (That complainant had subsequently been fired by Exxon and was at the time of the Valdez disaster suing in federal court for wrongful termination. This court case made information about the captain's pattern of alcohol abuse immediately available.) The observation that someone abusing alcohol is not an appropriate person to command a large oil tanker did not require any special expertise or quantification of the harm that he might cause.

In contrast, in all likelihood some sort of cost-benefit calculation (a short-sighted one) underlay the decision that it was not worth the expense to make the Valdez (and other Exxon oil tankers) double-hulled. (The spill has now cost Exxon over a billion dollars.)

Responsible Behavior in Assessing Risk

Cost-benefit and risk-benefit calculations are frequently made as a part of environmental impact statements. Although one might wish to completely eliminate accidents such as major oil spills, that would be possible only if transportation of oil were eliminated or severe restriction were placed on the size of tankers. The lesson from the catastrophic effects of the Valdez spill for those who conduct cost-benefit or risk-benefit calculations with environmental implications is to consider what makes for a morally responsible use of such techniques.

Analysis of environmental risks is often some engineer's official responsibility, as well as something that any engineer may find necessary to fill her professional moral responsibility for safety. Such analyses are a regular part of the assigned work of many environmental engineers and chemical engineers. As we saw in the introduction on concepts, official obligations and responsibilities differ from moral ones in both their ethical significance and their functioning. There is no guarantee that what an official obligation requires one to do is even ethically permissible. If it is, and one freely takes on the job or assignment, then there is some moral obligation to keep one's implicit promise to perform the job or task. Given an ethically permissible job assignment or official responsibility, the ethical question is how to carry it out in an ethically responsible way. Carrying out moral responsibilities, or official responsibilities in a morally responsible manner, have both ethical and technical aspects.

The minimal requirements for morally responsible behavior in engineering and science are competence and honesty. Honesty requires more than not falsifying information, of course; it also requires a balanced assessment and presentation of the situation. The requirement for a balanced presentation by engineers and scientists contrasts with the expectation on members of the legal profession who act in an adversarial setting as an advocate for one side or the other. Stephen Unger emphasizes this point in the case of Morris Baslow. Baslow was a marine biologist who investigated the effects of once-through use of waters of New York's Hudson River for cooling of electrical power plants. Baslow found a pattern of fish kill and inhibited growth rate of fish due to increased water temperature. His employer, LMS engineers, in deference to the interests of their client, Consolidated Edison, tried to suppress his results and continued to present a contrary view at hearings held by the EPA and the Federal Energy Commission. Baslow then told them that if they would not release his data, he would. He sent his letter to the EPA and was fired on the same day in 1979. He had insisted on a full and balanced reporting of the data, but LMS had wanted to release only that information that was in the interests of their client. [11]

The requirement for a full and balanced reporting of the facts needs emphasis in the context of environmental analyses, because of the cynical perception that environmental assessments are usually one-sided. There certainly has been some clear dishonesty in these matters. For a long while one of my colleagues who works in the area of risk assessment had posted on his office door an advertisement from a firm doing environmental assessments. It promised to do an environmental study "favorable to you," a clear offer to skew research to suit the client's interests.

Some sources of bias in research are illimitable. One of these is disciplinary bias. Researchers use the tools of their discipline and not of another; they cannot be expected to even know about all the methods from other disciplines that might be used in the situation. Although such bias can distort findings, it does not represent moral failings in a researcher.

A special challenge in rapidly developing fields is to keep up with new knowledge and techniques. In areas such as environmental engineering or biomedical engineering that have immediate implications for well-being, the need for an overview of alternative methods is especially pressing.

ECOLOGICAL THINKING AND THE QUESTION OF
WHO COUNTS

What is the justification for the growing opinion noted earlier that engineers have a professional responsibility to protect the environment? Why should anyone, engineer or otherwise, be said to have an ethical responsibility for the environment? Most of the environmental problems for which engineering solutions have been sought involve threats to human health and safety. These are also some of the most widely discussed threats to the environment; indeed, the threat they pose to human health may be the reason that they receive the attention they do. When environmental damage threatens human health and safety, the responsibility to protect the environment against that damage simply follows from the professional responsibility for public health and safety.

Environmental damage that does not directly threaten human health and safety often threatens some other aspect of human well-being. For example, the global warming caused by increases of "greenhouse gasses" threatens agriculture and hence economic well-being, as well as producing a secondary threat of physical injury to humans from droughts, coastal flooding, and storms. Such examples suggest that concern for the environment may derive solely from a concern about the effect that environmental destruction might have on humans.

Sometimes damage to the environment poses no clear threat of injury or disease to humans but does threaten some endangered species or a fragile ecosystem. Even in these cases, the possible loss of benefits to humans is often mentioned. For example, the extinction of species caused by the destruction of the Amazonian rain forest is hypothesized to eliminate potential sources of medical remedies. What is believed to be the last remaining sample of the smallpox virus (a virus that normally cannot survive without a living host) has been artificially maintained. This sample has been preserved, not out of concern for the smallpox virus, but because the sample might provide useful information some time in the future. Is resulting harm to humans a necessary condition for environmental damage to be ethically objectionable or for professionals with relevant knowledge to have a moral responsibility to seek to prevent it?

Moral Standing and the Environment

Those who hold that human use of the environment has limits other than what is needed to prevent harm to humans often argue by analogy with the plight of previously disenfranchised people. They argue that in times past, noncitizens, women, and children lacked some legal rights that we regard as basic and as matched by moral rights, many of them human rights. They argue that we do not now think of the "natural environment" as having legal rights only because of the immaturity of present-day ethical reflection. To understand and evaluate such arguments, we need to understand what sorts of objects are held to have moral standing, what that moral standing entails (for example, does it entail having moral "rights"?) and why we should think that those sorts of objects do have that moral standing.

Some of the most quoted arguments are rather unclear about the first point. For example, in an oft-quoted essay, "Should Trees Have Standing?," Christopher Stone proposes that legal rights be accorded trees, forests, oceans, rivers, and other "natural objects" – the natural environment as a whole.[12] He equates "having standing" with "having rights" and ascribes rights, although not the same rights as humans, to all of these beings. As we saw in the introduction, some people object to ascribing moral and legal rights to beings that cannot choose when they wish to exercise them. To say that such beings have moral standing is not open to the same objections. Having standing would also morally constrain actions toward such beings, so Stone's sort of argument will be examined as an argument about moral standing.

If we focus attention on individual organisms or even species, the concern is likely to come out as the implausible assertion that either every organism or every species has moral standing. Do viruses or at least species of virus have moral standing? Do all ticks or all species of tick? However, to ask this sort of question about individuals, even about individual species, is to ignore the ecological perspective. That perspective focuses on ecosystems. The resilience or robustness of the whole depends on the multiplicity of species that fill each "niche," that is, the *unimportance* of an individual species to the survival of the system.

> It may be difficult to draw the boundaries between one organism and another, let alone distinguish the good of one from that of other organisms. For example, we dwell in intimate interdependence with the bacteria that live within us. According to biologist Lynn Margulis, the normal adult human is about 15% bacteria by dry weight.

Species diversity makes an ecosystem more resilient because assaults to one species, say disease, are more likely to be able to be compensated by other species. However, this does not imply that each species (or indeed each feature of the soil or atmosphere) within each ecosystem has a claim to ethical consideration. If each species is entitled to such consideration, it must be for other reasons.

The moral standing argument applied to the environment is not as clear as it is when applied to individuals of other species. The systems-thinking characteristic of environmentalism fits poorly with debates about the moral standing of individuals. Since systems are more than the sum of their components, saying that ecosystems have moral standing elevates their preservation or well-being even at the expense of the individuals in them. If the flourishing of one group of individuals threatens to throw the system dangerously out of homeostasis, concern for the ecosystem would dictate that those individuals or species would better be sacrificed. If the flourishing of the species is considered primary, the suffering or survival of individual members of that species becomes less important. This tension may partially explain why, as Aarne Vesilind has noted, there is not much affinity between the ecology movement and the animal rights movement.

Saying that ecosystems do not have moral standing also leads to paradoxes, however. If ecosystems lack moral standing, how can there be any reason to preserve them except to benefit humans? [13]

A much-discussed case is that of the threat to the habitat of the spotted owl caused by destruction of forests. This case is often represented as a polarized choice between preserving the endangered owl and protecting the economic interests of the timbering industry along with those whose livelihood is bound up with the timbering interests.

The Case of Timbering and the Northern Spotted Owl

The northern spotted owl has for hundreds of years lived in the "old-growth" forests of the Pacific Northwest. It feeds on the plant and animal life created by decaying timber. Its habitat has dwindled over the past one hundred and fifty years as a result of heavy logging in the area, much of it on public land. An estimated two thousand pairs of spotted owls are all that survive today.

In 1986, environmentalists petitioned the U.S. Fish and Wildlife Service to include the spotted owl among "endangered species." The petition was met with staunch resistance from the timber industry in the region, since it would bar them from clearing the forests that are the owl's habitat. In 1990 the owl was finally declared a "threatened species." Timber companies were required to leave at least 40% of the old-growth forests within a radius of 1.3 miles of any spotted owl nest or site of activity. The timber industry claimed that this requirement would throw thousands of loggers and mill workers in the area out of work. [14]

The spotted owl conflict is frequently used to illustrate supposed conflicts between, on the one hand, the ecosystem for which the spotted owl is an indicator, and on the other, almost thirty thousand jobs. The environmentalists point to both the aesthetic value of this ecosystem and its scientific value, that it has taken millennia to create and would not be restored by reforestation. Those concerned with job loss point to the host of social ills, from domestic violence to suicide, that regularly attend job loss. In this way, the question of environmental responsibility is frequently presented as a debate between two sides, one advocating environmental protection and the other advocating job preservation.

As portrayed, the spotted owl case may not be representative of the underlying trade-offs involved in environmental protection, but rather only reflect common conflicts among interested individuals and groups. That is to say, there may not in most cases be a need to trade off environmental preservation and job creation, even in the short run. (In thirty years, at the current rate of timbering, the old growth forests would be gone, forcing the mills to close.) A recent study showed that the Endangered Species Act has not had a negative economic impact at the state level. In fact, states with booming economies were found to be the ones that also had the largest number of federally listed species, contradicting the impression that the Endangered Species Act is creating major economic harm. The underlying mechanism for the observed findings seems to be that population growth goes with economic boom and tends to put greater pressure on the environment, leading to new listings of endangered species. The study does not deny the economic effects of environmental preservation, but it finds these to be highly localized, such that they are "lost in the noise of background economic fluctuations." The study cites for comparison the recent series of military base closings, and finds the economic effects of those closings to be hundreds of times greater than the combined effects of all the listings under the twenty-year history of the Endangered Species Act.[15] The human problems of job loss and dislocation are real and require a more adequate societal response, but environmental protection is not a principal source of these problems.

As we saw in the first chapter, responsible behavior requires careful consideration of the effects of one's actions. In thinking about the protection of the environment we are still developing ways of understanding the situation we face. The problem gets distorted when it is represented as a dilemma forcing a choice between economic hardship and environmental degradation. It may not be necessary, and it certainly is not sufficient, to permanently sacrifice jobs to protect the environment.

The 1995 Supreme Court Decision on "Taking" of a Threatened Species

As we consider what actions will be responsible, quite specific limits have to be set on what individuals and groups may or may not do. The Supreme Court in its decision on Babbitt, Secretary of Interior, et al. v. Sweet Home Chapter of Communities for a Great Oregon on June 29, 1995, addressed this very question.

The Endangered Species Act of 1973 made it unlawful for any person to take endangered or threatened species and defines "take" to mean to "harass, harm, pursue," "wound," or "kill." Secretary of the Interior Bruce Babbitt brought this suit. He interpreted "taking" to include actions that so significantly modified wildlife habitat as to kill or injure protected species. The respondents claimed that Congress did not intend the word "take" to include habitat modification. The District Court had found for Babbitt, but the Court of Appeals ultimately reversed the District Court decision. The Court of Appeals held that "harm," like the other words in the definition of "take," should be read as applying only to the perpetrator's direct application of force against the animal taken. In a six-to-three decision, the Supreme Court held that Babbitt's interpretation of Congress's intent was reasonable. [16]

For the reasons discussed earlier, we will set aside the question of whether the prohibition against "taking" members of endangered species finds its ultimate justification in the moral standing of a species, in the public interest in diversity of species, or in something else. Now we shall turn to another instance in which environmental damage came from an unsuspecting source. This is another case that has taught us that some species and ecosystems are extremely fragile; human actions that disturb the environment may have many far-reaching effects that we are just beginning to appreciate. It is not so much that we know our actions will have certain consequences that will threaten or damage other beings and that they along with ourselves have moral standing; rather, disturbing the environment is now recognized to frequently have major unpredictable consequences. The threat of those consequences gives us reason to question whether it is wise, prudent, or responsible to go forward. There is increasing appreciation of the danger that unintended consequences may do major and irreversible harm before they are detected or well understood.

An excellent example is the threat of global warming due to an increase in greenhouse gases. The year 1995 was the warmest on record, according to both the British Meterologic Service and the NASA Goddard Institute for Space Studies. The second warmest year was 1990, and according to British figures, the period 1991–1995 was warmer than any other five-year period, including 1980–1984 and 1985–1989, even though the 1980s had been the hottest decade on record.[17] It is possible that this warming is due to climate variations that have nothing to do with the increase in greenhouse gases, as those who have warned about this phenomenon acknowledge. However, their argument is that we cannot wait until we are certain that greenhouse gases are a principal cause, because at that point it would be too late to forestall devastating effects.

ACID RAIN AND UNFORESEEN CONSEQUENCES OF HUMAN ACTION

An ironic example of environmental damage from unforeseen consequences was the worsening of the problem of acid rain that resulted from a provision of the 1970 Clean Air Act. To reduce local air pollution in conformity with the requirements of the Clean Air Act, utilities and smelters built taller smokestacks.

The burning of fossil fuels, principally by coal- and oil-burning power plants and by automobiles, puts sulfur dioxide and nitrogen oxides into the air. What policy makers had not foreseen was that taller stacks would allow particles and gases to be carried farther and that the resulting sulfur dioxide and nitrogen oxides in the atmosphere would produce acid rain. The result of the taller smokestacks was to worsen the problem of acid rain in New England, the Adirondacks, Appalachia, and Canada.

Acid rain can damage forests and soil, degrade ecosystems, and kill fish, as well as damage buildings and statues. This acid rain has made about 1,000 lakes in the United States chronically acidic and has made over 14,000 Canadian lakes so acid that fish can no longer live in them.

The Clean Air Act was amended in 1990 and further amendments are currently being considered. Time will tell whether these amendments will have their intended effect.

The example of the negative effects of an act intended to protect the environment illustrates that environmental protection involves more than preventing dumping and other wrongdoing. Science and engineering knowledge is essential to the prevention, or at least the surveillance and early detection, of damaging consequences of well-intended and seemingly innocent actions. The new disciplines of environmental engineering and environmental science have arisen to answer the need for such expertise, along with that in emergency planning, waste separation and management, waste reduction, and the cleanup of hazardous chemicals and radioactive wastes.

The environment is sometimes discussed with a romanticism that makes it seem that all that is necessary is a different view of or attitude toward nature. Science and engineering are even occasionally represented as the enemies of nature, rather than the disciplines that study nature. More balanced accounts acknowledge that social practices and innovations of many types caused environmental damage well before the age of science. It was, after all, agrarian livestock damage to common land that gave rise to the expression "the tragedy of the commons." This expression refers to the tendency of individuals to seek their narrow self-interest, even when this leads to harm for everyone in the long run.

Important as it may be to recognize environmental responsibility as a responsibility sanctioned by religious and cultural traditions, viewing nature as sacred did not by itself prevent the prescientific societies from hunting some species to extinction. Responsibility requires knowledge, as well as concern, for its fulfillment. Just as one can endanger the public safety through ignorance as much as through evil intent or recklessness, so people can and do endanger either the environment or human well-being by acting beyond their competence. Most of the knowledge required for environmentally responsible behavior is engineering knowledge and scientific knowledge.

EFFECTS OF CHLOROFLUOROCARBONS ON THE OZONE LAYER

Another dramatic example of a major unintended consequence of human action is the depletion of the ozone layer by the use of chlorofluorocarbons. Ozone (O_3) is a powerful oxidizer and has a dual effect on the environment. As pollutant in the lower atmosphere it has greatly increased in recent decades and is a health hazard to humans and is harmful to crops. The ozone layer in the stratosphere nine to thirty-one miles above the Earth protects humans and other living organisms from the ultraviolet radiation of the sun. There ozone, together with oxygen (O_2) blocks a major part of that radiation. The actual number of ozone molecules comprising the layers is not very large; if the ozone layer were at standard temperature and pressure it would be as thin as a piece of cardboard. Therefore, a relatively modest

254 ETHICS IN ENGINEERING PRACTICE AND RESEARCH

amount of chemical pollutants can have a significant effect on the ozone layer. [18www]

Ozone is made in the upper atmosphere by the splitting of O_2 molecules into atoms of oxygen. These combine with other O_2 molecules in the presence of other air molecules to form ozone. As was demonstrated in part by Paul Crutzen in 1970, the decomposition of ozone into oxygen is enhanced by the presence of hydrogen radicals OH and HO_2, nitrogen oxides NO and NO_2, and free chlorine atoms. [19]

Mario Molina was a postdoctoral fellow in the laboratory of Sherwood Rowland at the University of California at Irvine in 1974 when they published an article in *Nature* on the threat to the ozone layer from chlorofluorocarbon (CFC) gases, or "freons." They built on the work of James Lovelock, who had developed the electron capture detector, a device that was able to measure extremely low organic gas contents in the atmosphere. With this device Lovelock had shown that CFC gases had already spread globally throughout the atmosphere. Molina and Rowland argued that CFCs could be transported up to the ozone layer where the intensive ultraviolet light would cause chemical decomposition, releasing chlorine atoms that would deplete the ozone layer.

Molina and Rowland calculated that continued use of CFC gases would deplete the ozone layer markedly in a few decades. Their calculations drew much criticism, but also much concern. Their work has now proven to be essentially right and even to have somewhat underestimated the risk.

The use of chlorofluorocarbons started out quite innocently. The stability and nontoxicity of these manufactured chemicals had made them seem safe and environmentally benign. They found many uses as refrigerants, industrial cleaning agents, propellants in aerosol sprays, and blowing agents in plastic foams; it was extremely unwelcome news that the use of CFCs was causing major environmental destruction.

Chemical erosion of the ozone layer has already resulted in dramatically increased rates of skin cancer, particularly in places in the Southern Hemisphere.

Paul Crutzen, Mario Molina, and Sherwood Rowland jointly received the 1995 Nobel prize in chemistry for their pioneering work in explaining how ozone is formed and decomposes through chemical processes in the atmosphere. These three researchers were honored for what the Nobel Foundation cited as contributing "to our salvation from a global environmental problem that could have catastrophic consequences." [20] This was the first Nobel prize to be given for environmental science.

SUPERFUND SITES AND THE MONITORING
OF COMMUNITIES FOR TOXIC CONTAMINATION

Although mistakes by well-intentioned people are the source of some of the current problem of toxic contamination, negligent, reckless, or even deceitful actions have also contributed to it. Particularly serious contamination with both radioactive and toxic substances has occurred at military installations. The cost of the cleanup of plutonium contamination alone is conservatively estimated at $150 billion, and the cleanup of all contaminants may cost $400 billion. That is more than the cost of the entire interstate highway system in the United States. The enormous contamination was allowed to worsen over many decades, because the Pentagon, the Energy Department, NASA, and the Coast Guard regarded pollution on their property to be none of the public's business. Leaders feared both embarrassment from disclosure and that cleanup would distract them from what they considered more important problems. Thus, although the 1970s environmental groups, state agencies, and the Federal EPA complained about toxic contamination, their warnings mostly went unheeded. The cleanup program now necessary has been described as "one of the biggest engineering projects ever undertaken." The Hanford Nuclear Reservation in Richland, Washington is the most extensively contaminated with both radioactive wastes (including plutonium and other radioactive elements) and chemical contaminants. These substances have contaminated the groundwater and soil and are seeping into the huge Columbia River that forms the western part of the boarder between Oregon and Washington.[21]

One of the most famous cases of toxic contamination by a corporation is that of Love Canal. In the 1890s, William T. Love had dug a trench as part of a plan to provide hydroelectric power for a model city he proposed to construct near the city of Niagara Falls, New York. Soon a depression left Love with no investors and his project was abandoned, and for the early decades of the twentieth century the trench was used as a swimming hole.

Beginning in the 1930s, chemical companies that had moved into the area began using the canal as a waste dump. In 1942, Hooker Chemical and Plastics Corporation negotiated an agreement to dump wastes at the site and eventually purchased the land for that purpose. They lined the site with cement to keep the chemicals from leaking. Over the next eleven years they put an estimated 352 million pounds of chemical waste in the canal.[22] In 1953, when the canal became full with chemical wastes and the municipal wastes that the city of Niagara Falls also dumped at the site, Hooker covered it with a clay cap.

The Niagara Falls Board of Education became interested in obtaining the land for a new school. Hooker Chemical took samples to demonstrate to the Board the presence of the chemicals at the site, but the Board was not dissuaded. Hooker finally sold the land by selling it to the Board for the nominal price of one dollar.

The Board of Education constructed a grade school and playground near the center of the parcel, adjacent to the landfill. In doing so they partially removed the cap that Hooker had placed on the site. Later the city punctured the cement liner when they installed new sewers and storm drains in the area. As demand for housing increased, the Board of Education sold the remaining land to developers who subdivided it for single family homes. During the next two decades the waste migrated through the soil, contaminating storm sewers and basements, or surfacing to evaporate and contaminate the air. Beginning in 1958, residents complained to the city about foul odors, oily liquids in their basements, and rashes on children who attended school or played at the site.

Environmental monitoring at the Love Canal was first conducted in 1976. Data collected by the New York Department of Environmental Conservation (NYDEC) and by Calspan Corporation, a private firm hired by the city, revealed extensive contamination of the groundwater, soil, and air. When the local newspaper published these results, the frustrated citizenry took the matter to their Congressperson, who involved the federal Environmental Protection Agency (USEPA).

The New York State Health Commissioner received a report on the site from the USEPA in March, 1978. Three months later the State Department of Health conducted a house-to-house health survey. The health department drew blood samples from those living in the ninety-seven homes immediately adjacent to the canal. Two days later Governor Carey declared a state health emergency.

Various citizens' groups emerged in response to the crisis. One such group, under the leadership of Lois Gibbs, later evolved into the Citizens' Clearinghouse for Hazardous Wastes, a citizen's group formed to help other communities in similar circumstances. [23]

This case was discussed extensively in 1979 Congressional hearings that led to the passage of the Comprehensive Emergency Response Compensation and Liability Act (CERCLA), popularly known as "the Superfund Act." Health studies and litigation continued. A settlement was reached between former residents and Occidental Chemical, the parent company of Hooker, in 1983. A new clay cap was installed over the canal in 1984. A consent agreement was reached between the U.S. government and Occidental Chemical in 1989.[24]

Although many sites of toxic contamination exist, the story of Love Canal was a landmark case and one of the best documented, because it was the first to receive national attention. It should be recognized, however, that there are worse sites in the United States. A 1987 study by the United Church of Christ Commission for Racial Justice found that three out of five African-American and Hispanic Americans live in communities with uncontrolled hazardous waste sites. As Robert Bullard documents in *Dumping in Dixie, Race, Class and Environmental Quality*, many of the worst sites that have received less attention are in the South and primarily affect the health of people of color.[25]

The threat of toxic contamination will be with us for the foreseeable future in view of the large and increasing number of known toxic waste sites and the frequency of acute chemical "releases" (spills and the like). More than 30,000 hazardous waste sites are known to exist in the United States. Of these, more than 1,200 are large enough or affect large enough populations to be listed on the Environmental Protection Agency's National Priorities List of Superfund cleanup sites. The inventory of Superfund sites has been growing each year.[26] The inventory of *potential* Superfund sites is growing at the rate of two or three thousand annually.[27]

Along with toxic exposure from dumps, toxic exposures due to chemical releases are frequent. There are about 1,200 acute chemical releases in a typical year. In 1986, accidental releases resulted in 210 deaths, 6,490 injuries, and 533 evacuations.[28] Engineers and scientists are necessarily involved in investigating these sites, releases, and accidents, attempting to prevent or mitigate the effects of future accidents and dumping and to clean up what has already occurred.

The great expense of the needed cleanup, especially the enormous expense of cleaning up plutonium at military installations, makes this a responsibility that citizens and public officials are tempted to ignore. Such inaction would leave the toxic and radioactive materials to continue to seep into the soil and poison groundwater. The need for responsible action by many groups, institutions, and nations shows that corporate attitudes are not the only ones that need improvement. However, since many engineers and scientists work in corporate environments, understanding corporate attitudes and the variety among them is particularly important for engineers and scientists, especially those at the beginning of their careers.

CHANGING NORMS IN U.S. CORPORATIONS

Attitudes about pollution are changing rapidly in U.S. corporations. How much has changed in the last ten to fifteen years is highlighted by a study of corporate attitudes conducted by Joseph M. Petulla from 1982 to 1985.[29] Even during the three years of his study, Petulla noted increasing environmental concern. He rated companies in one of three categories based on their management practices regarding environmental pollution. Contrary to his expectation, Petulla did not find a direct correlation between these practices and company size. He did find that the "corporate culture" as determined by the CEO, and, in some cases, the senior corporate attorney, was a major factor in predicting a company's management strategy.

The most environmentally concerned companies demonstrated what Petulla characterized as "supportive compliance" with environmental laws. Compliance was endorsed by the CEO, carried through by well-trained personnel using the best pollution control equipment, and supported by on-going research and cooperation with government agencies and community groups. However, Petulla found that only 9% of the companies surveyed fell into this category. The middle category, into which 58% of his sample fell, demonstrated "cost-oriented environmental management"; that is, they complied with the law but demonstrated no general commitment to preventing environmental degradation. The final group, into which 29% of Petulla's sample fell, was classified as demonstrating "crisis-oriented environmental management." They had no full-time staff assigned to environmental protection and addressed issues of pollution only when forced to, frequently finding it cheaper to pay fines and lobby than to prevent pollution. One representative from this group expressed his reason for this strategy by saying, "Why the hell should we cooperate with the government or anyone else who takes us away from our primary goal [of making money]?"[30]

The best known statement of the position that making money is the primary goal of corporations is the one mentioned at the beginning of Chapter 5 by economist Milton Friedman. He said that the responsibility of managers is to "make as much money as possible while conforming to the basic rules of society, both those embodied in law and those embodied in ethical custom." Friedman specifically says that stockholders' money should not be spent on "avoiding pollution."[31] He advocates obeying the law, but his view is that it is the responsibility of a manager to do no more than legally required and even to legally resist compliance where doing so will make more money.

Some companies began initiatives in environmental protection somewhat earlier, but many of them only became significant in the 1980s. The aim of such programs is not merely to reduce pollution from waste. They also often aim at changing manufacturing methods so as not to use hazardous substances and not to produce them or pass them on in the product or byproducts.

The 1990s have seen a rise in the popular support for control of environmental pollution by groups ranging from the Green Party in California to the Religious Right. In this changing climate, many corporations now at least wish to appear environmentally concerned. This has been particularly true of chemical companies.

Nonetheless, even with greater corporate commitment to environmental protection, there is still a range of corporate attitudes. Experienced engineers have described to me great differences even within the same industry. It is important for new engineers to know what sort of company they are getting into.

SUMMARY AND CONCLUSION

We have seen how new awareness of environmental degradation and an ecological understanding have developed since the early 1960s. New disciplines of environmental engineering and science give powerful evidence of the relevance of these disciplines to environmental protection. A variety of reasons are given for thinking that engineers have a special professional responsibility for the environment. The environmental effects that have received the bulk of attention have been the ones that carry clear implications for human health and safety. Such effects fall under the well-recognized responsibility of engineers to ensure the public health and safety. Some environmental damage does not have clear implications for human health and safety, however. Should responsibility for such environmental effects be viewed as on the model of the engineer's responsibility for quality, perhaps? Or is there no reason for thinking that engineers have any special responsibility to prevent such damage? (All people might have some responsibility, or particular engineers, such as environmental engineers, might have responsibilities stemming from their position of trust.) There is not a clear answer, but we may not need one. Ecological thinking leads to the recognition of the interconnectedness of organism and species and the recognition that the good of one is often highly dependent on others or on the whole. Therefore, the question of the moral standing of individuals and species may be of little importance so long as some interdependent beings have moral standing. Recent examples such as the increase in acid rain, depletion of the ozone layer, and the toxic contamination of groundwater have taught us that such damage can have far-reaching negative effects on the environment, effects that clearly endanger human health, safety, and well-being. It is prudent to consider that any major environmental effect may turn out to harm some individuals with moral standing. Engineering and applied science are essential to help anticipate the consequences of human action on the environment before the effects become catastrophic.

The attitudes of U.S. Corporations toward the environment vary widely, although environmentalism in corporate America has increased significantly in the past two decades. Because of this difference in corporate attitudes, engineers will find a variety of levels of support for fulfilling their newly recognized responsibility for the environment.

NOTES

1. Adapted from 92-6 of the Board of Ethical Review of the NSPE. The NSPE cases are also available in hard copy in volumes V, VI, and VII *of Opinions of the Board of Ethical Review*, Alexandria VA: National Society of Professional Engineers.
2. *Gilbane Gold* is a 24-minute videotaped dramatization produced by The National Institute for Engineering Ethics, National Society of Professional Engineers, 1989. That dramatization raises many ethical issues, including how to cope with such a flaw in environmental legislation.
3. Merchant, Carolyn. 1980. *The Death of Nature: Women, Ecology, and the Scientific Revolution.* San Francisco: Harper & Row.
4. Briggs, Shirley A. 1987. "Rachel Carson: Her Vision and Her Legacy," *Silent Spring Revisited*, pp. 4–5.
5. Quoted in Briggs, 1987, p. 5.
6. PBS, *The American Experience: Rachel Carson's Silent Spring*, and *Encyclopedia Britannica* online, http://www.eb.com/.
7. *Ibid.*
8. The Delaney Clause in Section 409 of the 1958 Federal Food, Drug and Cosmetic Act (FFDCA) of 1958 was the only national legislation in place prior to the publication of *Silent Spring*. That clause pertains to "food additives" and specifies that no amount of carcinogenic pesticide residues is to be tolerated in processed foods.
9. I have placed this list in the WWW Ethics Center for Engineering and Science as supplementary material to the story of Rachel Carson. There the list is linked to further information about the acts and agencies in charge of them.
10. Vesilind, P. Aarne, J. Jeffrey Peirce, and Ruth Weiner. 1987. *Environmental Engineering*, 2nd edition. Stoneham MA: Butterworth. pp. 6–7. Discussion of concepts of environmental engineering in this chapter derives principally from Vesilind et al.

11. Unger, Stephen H. 1994. *Controlling Technology: Ethics and the Responsible Engineer.* Second edition. New York: John Wiley & Sons, Inc.
12. Similar features are found in more recent influential arguments as well. See, for example, Midgley, Mary. 1992. "Is the Biosphere a Luxury?" *Hastings Center Report*, 2(2):7–12.
13. Refer to the discussion of moral standing in Part 2 of the introduction on concepts.
14. The account to this point is based on "Ethics and the Spotted Owl Controversy." *Issues in Ethics*, IV(1) (Winter/Spring, 1991):1, 6.
15. Working paper "Endangered Species and State Economic Performance" by Stephen M. Meyer, published by the Project on Environmental Politics and Policy at MIT, reported in "Study Finds Small Economic Effects from Endangered Species Protection," by Robert C. Di Iorio. MIT Tech Talk, 39(26) (April 12, 1995):1, 8.
16. This account is based on the summary ("syllabus") prepared by the Supreme Court's Reporter of Decisions and distributed by e-mail on the list <liibulletin@fatty.law.cornell.edu>.
17. Stevens, William K. 1996. "1995 the Hottest Year Recorded on Earth," *New York Times*, January 4, p. 1.
18. See the Nobel announcement on the 1995 prize for Chemistry at http://www.nobel.se/announcement95-chemistry.html.
19. *Ibid.*
20. *Ibid.*
21. Schneider, Keith. 1991."Military Has New Strategic Goal In Cleanup of Vast Toxic Waste." *New York Times*, August 5, A1, D3. Extensive information on the WWW about the Hanford site may be found at http://www.halcyon.com/tomcgap/www/hanford.html.
22. Worobec, M. 1980. "An Analysis of the Resource Conservation and Recovery Act." *BNA Government Reporter Special Report* (August 22).
23. Gibbs, Lois M. 1985. *Centers for Disease Control: Cover-up, Deceit and Confusion.* Arlington, VA: Citizens' Clearinghouse for Hazardous Wastes; and 1982. *Love Canal: My Story.* Albany, NY: State University of New York Press.
24. The account here is drawn from a longer account in *Monitoring the Community for Exposure and Disease: Scientific, Legal, and Ethical Considerations* by Nicholas A. Ashford, Carla Bregman, Dale B. Hattis, Abyd Karmali, Christine Schabacker, Linda-Jo Schierow, and Caroline Whitbeck, a report supported by the Agency for Toxic Substance and Disease Registry (ATSDR) and the National Institute for Occupational Safety and Health (NIOSH), 1991. That account makes extensive use of the transcript of the hearings held by the U.S. House of Representatives (U.S. Congress 1979a) as well as the specific sources cited above. For a valuable insight into the thinking of managers within the Hooker Corporation, see Ch. 2.4, "The Hooker Memos" in Alastair Gunn and P. Aarne Vesilind (1986). *Environmental Ethics for Engineers.* Chelsea MI: Lewis Publishers.
25. Bullard, R. D. 1990. *Dumping in Dixie: Race, Class and Environmental Quality.* Boulder, CO: Westview Press.
26. Commentary by Charles Xintaras at the 1993 Hazardous Waste Conference on

Education in Environmental Health, online at http://atsdr1.atsdr.cdc.gov:8080/cx12a.html.

27. Zuras, A. D., F. J. Prinznar, and C. S. Parrish. 1985. "The National Priorities List Process." in *Management of Uncontrolled Hazardous Waste Sites*, ed. AIChE. New York: AIChE. pp. 1–3.

28. Binder , S. and S. Bonzo. 1989. "Letter to the Editor." *American Journal of Public Health*, 79(12):1681.

29. Petulla, Joseph M. 1989. "Environmental Management in Industry." in *Ethics and Risk Management in Engineering*, edited by Albert Flores. Landam, MD: University Press of America. This study was brought to my attention by reading Charles E. Harris, "Manufacturers and the Environment: Three Alternative Views" in *Environmentally Conscious Managing: Recent Advances*, edited by Mo Jamshidi, Mo Shahinpoor, and J. H. Mullins, Albuquerque: ECM Press, 1991.

30. Petulla, *op. cit.* p. 146.

31. Friedman, Milton. 1970. "The Social Responsibility of Business Is To Increase Its Profits." *The New York Times Magazine* (September 13); reprinted in *Ethical Issues in Engineering*, edited by Deborah Johnson, Englewood Cliffs, NJ: Prentice-Hall, Inc., 1991, pp. 78–83.

9

FAIR CREDIT IN RESEARCH
AND PUBLICATION

Jan and Keith are junior members of the engineering faculty at a major university. Both are seeking tenure from the university, and as part of the requirement, they are required to publish original articles in disciplinary journals.

Jan reviews some work from his work as a graduate student and reconsiders a paper based on an unpublished portion of his thesis research. He thinks that with some revision it would make a good journal article. Jan discusses this idea with Keith and proposes that together they revise the paper and bring it up to date.

Jan does most of the revising and updating. Keith makes only small contributions but is a better writer. Jan is disappointed that Keith does not make more of a contribution to the paper's content but agrees to include Keith's name as coauthor, to enhance Keith's chances of obtaining tenure. The article is accepted and published in a scientific journal.

Is it ethically acceptable for Jan to go back to his graduate student work for an article to publish?

Should Jan's thesis supervisor be credited in some way, and if so, how?

Should the source of the funding for Jan's thesis research be acknowledged in the paper?

Is it responsible for Jan to ask Keith to help revise the article? How much could they and should they have agreed upon at the start of their collaboration?

Is it either unethical or unwise for Jan to include Keith's name as an author? Is it unethical or unwise of Keith to be an author under these circumstances?

Credit for intellectual work is a hot topic. It is much discussed these days, and it has evoked strong feelings for centuries. Today new patterns of interdisciplinary collaboration and new forms of intellectual property, especially property that is stored in an electronically digitized form (which can be readily copied), complicate the question of how to apportion credit fairly.

As we saw in Chapter 6, plagiarism in a research context is misconduct, although plagiarism differs from acts like fabrication and falsification in that plagiarism does not alter the research itself. Questions of plagiarism or of fair credit for individuals, generally, make sense only in social contexts in which the names of individuals are publicly associated with particular research contributions. Even today, within industry individuals are usually credited for specific achievements only within their groups or companies. They receive other rewards, such as bonuses or promotion, in place of credit for the work itself. Organizing and rewarding work may occur without publicly crediting individuals for particular accomplishments. Whatever the system of rewards, if the reward system breaks down, the trust necessary to sustain the enterprise dwindles.

Plagiarism is the most clear-cut abuse of standards of fairness in assigning credit for individuals' publishable research in engineering and science. The following scenarios illustrate other more subtle abuses.

PLAGIARISM IN A GRANT PROPOSAL

You are making steady progress on research for the doctorate and have submitted a report on some interesting new experimental work to the large multidisciplinary laboratory in which you are doing your work. Another group in the same large laboratory is eager to get grants that will fund another student's thesis and is busily writing proposals. After one of these grants is funded, you discover that its authors had proposed to do exactly the work described in your report, a report they received. The writers of the new grant even incorporated the language of your report into their proposal.

What can and should you do?

SURPRISE AUTHORSHIP, CREDIT, AND RESPONSIBILITY[2]

As a beginning graduate student you worked in a research team with two other students and the supervising professor. One of those students, who was close to finishing the Ph.D. and anxious to build a professional reputation, sent off many abstracts proposing to present papers at meetings. This student graduates before most of these society meetings have occurred.

One day when you are looking through the proceedings of one of the professional societies to which the advanced student made a submission, you find that a paper has indeed been presented and that it has your name on it as coauthor.

The paper is in two parts. The first represents some group work in which you participated. It is only loosely related to the second part, which concerns a point of theory. Nonetheless, the conclusions in both parts of the paper appear respectable.

What, if anything, should you do?

UNCREDITED AUTHOR

Ames, head of the Mangrove Company, offers to provide funding to two faculty members of the chemistry department of a major university for research on removing poisonous heavy metals (copper, lead, nickel, zinc, chromium) from waste streams. Under the agreement the university agrees to give the Mangrove Company exclusive use of any resulting technology developed for water treatment and waste water stream treatment. Mangrove Company agrees to provide a royalty to the university from profits from the use of the technology and leaves the faculty members free to exploit applications of the technology other than the treatment of water and waste water.

While the university research is being conducted, Mangrove continues to conduct research in the same area. Performance figures and conclusions are developed. Mangrove freely shares the figures and conclusions with the two members of the chemistry faculty.

At the university, Bradley, a professor of civil engineering, wants to conduct research and write a paper on the use of the technology to treat sewage. Bradley contacted the two members of the chemistry department. The chemistry professors provided data (primarily Mangrove's) to Bradley for use in the research and paper. The chemistry professors did not tell Bradley that the data were obtained by Ames and the Mangrove company.

Bradley's paper is published in a major journal. Ames's data are displayed prominently in the paper and the work done at Mangrove constitutes a major portion of the journal article. The paper lists the two chemistry professors as authors as well as Bradley. No credit is given to Ames or Mangrove. After publication Ames informs Bradley about the actual source of the data and findings.

What can and should Bradley do? What can and should the two chemistry professors do?

– based on the NSPE Board of Engineering Review Case 92-7 [3]

This chapter examines norms and practices for crediting research and the justifications for them, together with common pitfalls and problems. To better understand how research has arrived at these practices, we will begin by briefly reviewing the development of modern practices of authorship.

THE HISTORY OF THE SIGNIFICANCE OF AUTHORSHIP IN SCIENTIFIC RESEARCH

In the seventeenth century the British Royal Society was formed to foster the growth of what, in English, is now called "scientific knowledge." The British Royal Society became an important influence on the methods and practices of modern science that were emerging. This elite group of experts continues to advise the British government on scientific matters.

Some original members of the British Royal Society were jealous of credit for discovery. Isaac Newton (1642–1727) especially wanted credit for being the first to make certain discoveries. (He was extremely concerned about his priority vis-à-vis Leibniz in discovering calculus.) Newton resorted to distributing his findings in anagrams (a reordering of the letters describing his discoveries into a new, usually nonsensical, string of words). When another person claimed the same discovery, Newton could unscramble his anagram to show that he had discovered it first. "Publication" in anagrams, however, did nothing to advance knowledge.

Another barrier to the dissemination of knowledge was the interest of some researchers in waiting until they felt they had the whole story and so could publish the definitive treatise on a subject. Newton took decades to complete his *Principia*, for example. To create an incentive for open publication of new findings, Henry Oldenberg, the secretary of the Royal Society, offered researchers prompt publication of their work in the *Philosophical Transactions* of the Society, together with the promise that the Society would stand behind the scientists' claims to be the first to discover the phenomena in question. Thus was born the convention that priority goes to the person who publishes first (or now, more precisely, the person who first *submits* the finding for publication), rather than the person who first makes the discovery.[4] Notice that these practices for awarding credit were devised to circumvent some very destructive effects of competition among researchers; the practices were not the source of that competition.

Researchers in seventeenth-century Western Europe were men of the upper classes who had the wealth to support their research. Elizabeth Potter[5] has shown how gender and class assumptions explicitly influenced judgments about whose word could be trusted, and so who was a reliable witness of experiments. To question a gentleman's data was tantamount to questioning his word and so might precipitate a duel.

Election to the British Royal Society was based on a mix of political and disciplinary considerations and excluded those whose work was very practical. Thus neither James Watt (1736–1819), Scottish-born engineer and inventor of the steam engine, nor Thomas Sydenham (1624–1689), the pioneer of modern disease classification, were elected.

Recognition by august bodies did not entirely settle questions of priority. Many notable national rivalries and disagreements exist over who should be credited with a particular discovery. For example, histories of science written in English credit James Prescott Joule, a British physicist, as the person who discovered the Principle of the Conservation of Energy as applied to heat.[6] French and German histories, however, often credit Julius Robert Mayer, a German physicist, with this discovery. (All authorities agree that both scientists made contributions to the mechanical theory of heat.) Today, though, disputes about priority are common not only between nations, or between rival laboratories, but even among collaborators or members of the same laboratory. In part this is due to the large role collaboration plays in contemporary scientific research and in part to the slow development of ethical norms to establish agreement on standards of fairness among today's extremely competitive researchers.

The rush to publish can lead to the presentation of premature claims, which retards the progress of knowledge.

PLAGIARISM: FROM COPYING TEXT TO APPROPRIATING SIGNIFICANT IDEAS

Like falsification, plagiarism is a fundamental betrayal of trust. As we saw in Chapter 6, to plagiarize is to appropriate another's ideas or writings and to represent them as one's own. Plagiarism differs from fabrication and falsification in that the charge of plagiarism makes sense only in certain social contexts, namely those in which the names of particular individuals are associated with particular contributions. That way of organizing and rewarding work is characteristic of individualistic societies. Research in engineering, science, and medicine could be organized quite differently and would still be research. It is easy to point to large-scale projects in other cultures that, like the cathedrals of the Middle Ages, were completed without recording the names of those who built them. In many areas today the conventions for credit and citation are different from those of academic research. For example, news items and articles written for nonscholarly publications do not use footnotes or formal citations. They seek to accurately quote sources, but primarily to establish who holds what views, not who is the originator of a particular idea unless the story is about an invention or discovery.

Even in technologically developed democracies, corporations evaluate and reward the work of those in their employ who conduct industrial research and testing by promoting them to larger responsibilities, giving them titles that reflect those responsibilities, and raising their salaries, rather than primarily through the mechanism of publication credit.

These examples illustrate that scientific research can, in principle, exist without the conventions of authorship and other forms of publication credit. Therefore, it may seem strange to include plagiarism with fabrication and falsification; one might argue that only fabrication and falsification corrupt the research record. Plagiarism, it might be held, operates only to apportion credit for the work but does not affect the work itself.

Although the assignment of credit to individuals in our social system through the laws and conventions regarding authorship, copyright, and patent might be replaced by a diffierent one, *some* system of rewards for work is needed. Although individual assignments of credit may be seen as separable from the accomplishment, the whole system of assigning credit is part of the organization of work in the first place. If that system breaks down, then the trust and cooperation necessary to support complex activities such as research also break down. Confidence in fairness is necessary for cooperation to thrive.

Plagiarism is the most common accusation of scientific misconduct, both in the experience of research universities, where it accounts for about two thirds of the complaints of research misconduct,[7] and in the reports of the office of the Inspector General of the National Science Foundation. This higher rate of reported plagiarism may be due to a greater likelihood that plagiarism will be detected, to a greater likelihood that once detected, someone will report it (a reviewer whose work has been plagiarized often is the one who reports it), or to a higher incidence of plagiarism as compared with falsification and fabrication.

Plagiarism fails to credit another's contribution. Credit is given in three different ways, appropriate in different circumstances: inclusion as an author of the article, listing of one's contribution to the work in a formal acknowledgment (usually in a footnote to the article or in an "Acknowledgments" section), or citation of a work in the "References" section of the paper. Commonly, works cited are previously published or publicly presented works, but reports internal to an organization and personal communications are also appropriate to cite.

Multiple instances of plagiarism have occurred on a variety of campuses. The following situations are typical:

1. Grant reviewers find that the grant proposal plagiarizes their own work or other work with which they are familiar.
2. Engineering faculty members do a literature search in their field and discover one of their own articles published, perhaps with the original diagrams, in another language under someone else's name.
3. Engineering graduate students helping their advisors review articles carefully read the references cited in the article to familiarize themselves with the topic and discover plagiarism of sources cited.
4. Two or more previous collaborators cease their collaboration and one uses some common work in a way that the other(s) take to be appropriating their ideas or work without credit.

Because the fourth type of situation involves collegial relationships, it is somewhat different from the first three. Some administrators have sought to classify it as something other than plagiarism, presumably because previous collaboration complicates the matter. Simply because the other person is a former collaborator is no reason, however, to say that the appropriation of another's work is not plagiarism.

In one extreme case a senior colleague plagiarized the work of a junior colleague without bothering to retype the manuscript. In 1985, Heidi Weissman, a specialist in nuclear medicine and radiology at Montefiore Medical Center in New York City, authored a chapter that was published with her name as sole author in a copyrighted volume published by the Radiological Society of America. The paper was on the subject of heptabiliary imaging, an imaging technique that Weissman had pioneered with the head of her laboratory at Montefiore, Leonard Freeman.

Freeman admitted twice substituting his name for Weissman's on the paper and presenting the work as his own without Weissman's knowledge or consent. The first time, he sent it to a publication in Taipei, Taiwan. On the second occasion he simply photocopied the text, removed Weissman's name with correction fluid and entered his own. He then distributed the paper as part of the course reading for a postgraduate course he was teaching.

In April 1987, several months before she discovered that Freeman had appropriated her work, Weissman filed a sex discrimination complaint against Montefiore Medical Center and the Albert Einstein School of Medicine. In August she discovered that Freeman had replaced her name with his and distributed it to his class. She sued for copyright infringement. The next day she was locked out of her office. In October she was fired and Freeman was promoted.[8]

Weissman reports[9] that when she sought redress through the American Association of University Professors, she was informed that the Albert Einstein School of Medicine was already under censure from the AAUP and had done nothing to have the censure removed. In view of the school's disregard of their disapproval, the AAUP did not expect to be able to influence them on Weissman's behalf. After some years Weissman was finally awarded a settlement totaling close to a million dollars.

The appropriation of Weissman's work by Freeman, taking every word of Weissman's paper, was such gross appropriation of another's work that it reportedly prompted Walter W. Stewart to propose that the extent of plagiarism be measured in units of a freeman, with a whole freeman being the maximum possible.

This is another example of a situation so badly handled that it showed a desperate need for better institutional mechanisms for handling charges of research misconduct. It is often cited in discussions of research ethics, but it is exceptional both in the flagrancy of the act and the irresponsibility shown by the institution, and in being a case in which a legal judgment of copyright violation could substitute for an ethical judgment of plagiarism. As we saw in the introduction, the legal notion of copyright has a somewhat different scope than the ethical notion of plagiarism. A copyright is a legal property right (usually of the author, composer, or publisher of a work) to the exclusive production, sale, and distribution of some work. Copyrights legally protect only "forms of expression," not underlying ideas. Plagiarism, in contrast, does include the appropriation of ideas, so copyright protection does not legally prohibit all forms of plagiarism. Plagiarism may be easier to *prove* when exact words, equations, or diagrams are copied, but misappropriation of credit for ideas without misappropriation of specific forms of expression still constitutes plagiarism. Plagiarism may be failure to give appropriate acknowledgment as well as a failure to include authors or to cite previously published work (although it is difficult to imagine a case in which failure to give an acknowledgment would be a copyright violation). In short, questions of plagiarism are not settled by determinations about copyright infringement, even when the work appropriated has been copyrighted.

THE CRITERIA FOR AND RESPONSIBILITIES OF AUTHORSHIP

The first significant consideration about authorship is who warrants inclusion as an author of an article. The second is the order in which the authors are listed. In some fields and journals, authors always appear in alphabetical order. Alphabetical listing is a necessity in fields like high-energy physics, where the number of authors reaches a hundred or more. In other fields and journals, the order of authors represents the importance of the contribution that each one has made to a paper. Thus, the **primary** or **"lead"** author is listed first.

In some other countries, however, the custom is or was for the **highest ranking** author's name to go first. This has been the practice in Germany and Japan, for example. (The highest ranking author is often called the "senior author," although this person need not be the oldest. However, the term is used in a variety of senses.[10]) In some fields (in the United States these are primarily medical fields) the practice is to place the name of the head of the laboratory last. The **last author position** then

Practices of authorship and credit within laboratories – especially laboratories in molecular biology and chemistry – together with gender issues, grantsmanship, and refereemanship, are knowledgeably presented by Carl Djerassi in his lively novel of research life, *Cantor's Dilemma*.[11]

Djerassi describes some of the comparative differences in the social organization of research laboratories between Japan and the United States in his second novel, *The Bourbaki Gambit*.[12]

has special significance and is reserved for the person who oversaw, coordinated, and probably obtained the funding for the research or for a larger enterprise of which the research was a part.

Another category of author, but one that carries no significance for credit for or role in the research, is that of **corresponding author**, the person to whom one writes for reprints of a multiply authored article. This person may have taken on this role for many reasons, such as being the only author who is not switching institutions or soon going on leave.

Two related questions should be answered in assessing the fairness of authorship assignment. First, what are the conventions about the significance of authorship and the order of authors? Second, is authorship credit, including any differential credit, implied by the ordering, fair? The present moral situation is one in which senior researchers and heads of laboratories who most often make decisions about authorship may often be relatively inarticulate about the criteria that they use. Some may even cover this inarticulateness with a refusal to state their criteria, perhaps with a show of outrage at being asked. Recently, while working on a statement on research ethics, I mentioned the interviews of present or potential research supervisors that the graduate students in MIT's computer science program and in some areas of electrical engineering conduct on policies and practices for assigning credit. One of the molecular biologists working on the statement said forcefully that he would refuse to discuss his criteria and practices with any student or post-doc, and tell that person if he wanted to work with him, he would just have to trust him. Opening the topic of practices for assigning credit to reasoned discussion, however, is an important step in establishing a basis for such a trust.

The specialization within science in this century has further complicated questions of authorship. Two physicists working in different areas are likely to know very little of the developments in each other's specialization. In a collaboration, therefore, they may be unable to evaluate each other's contribution. Research within a discipline may require an interdisciplinary team. Sophisticated data analysis may require the interpretation of a statistician, but the statistician may have very little understanding of the phenomena being investigated.

Interdisciplinary investigation has become common. Coauthors often understand little of their collaborators' work or even the problems the others address. For example chemical engineers and biologists may undertake collaboration on tissue culture or on scaling up some process for commercial use with little initial understanding of each others problems and methods. Such interdisciplinary collaboration complicates the assignment of credit, because with little shared understanding of the total effort, the researchers have difficulty evaluating the relative importance of each contribution.

To establish a clear understanding with their collaborators about how to conduct the research and how to share the credit and rewards for their research, engineers and applied scientists in interdisciplinary research must understand the differences in both the character of the research and the social organization of the research effort in all the disciplines involved. Although the consensus is that such understandings can prevent many difficulties and are best established early in the collaboration, negotiating them is not easy. Establishing such understanding may be even harder when participants come from different fields that have disparate journal practices, expectations for research, or understanding of the phenomena they jointly investigate.

IS THIS AN OUTRAGE OR A "CULTURAL" DIFFERENCE?

The problem that you are working on turns out to have significant applications for a new medical technology, so this year the funding for your graduate work has come from Dr. X's lab at one of the area's medical centers. You are writing up two papers for publication and show them to one of your coauthors, who tells you that Dr. X's name is always included on the author list in the last position. You object that Dr. X has not the remotest idea of what the research is about. You are told that, notwithstanding explicit statements to the contrary, this is often the practice in medicine and it certainly is "local custom" here.

What can or should you do?

The responsibilities of authorship received widespread attention in the famous case in which Thereza Imanishi-Kari was accused first of poor scientific work, and several years later of fabrication of experiments in research in which she had collaborated with Nobel Prize–winning biologist David Baltimore. In 1986, David Weaver, Baltimore's protégé, had published a paper with Imanishi-Kari and Baltimore in *Cell*. Although some press accounts of the case were grossly misleading and some parties incautiously spoke of "fraud" to mean fabrication after that charge was made against Imanishi-Kari,[13] the accusations against Baltimore were not that he had fabricated experiments, much less committed fraud. Rather, he was faulted for having his name on a paper with which he was not thoroughly familiar, and for trying to suppress the original university-level inquiry and the later congressional examination of the case to spare himself and David Weaver the embarrassment of a retraction[14] – a retraction in which he eventually did participate in 1991.

The case is an example of one that was made worse by poor complaint handling at the home institution. In their initial inquiry, MIT first failed to immediately obtain Imanishi-Kari's "notebooks" (more accurately described as lab records) as any research university now would.[15] Second, they did not promptly inform the complainant of their disposition of her complaint. The means of dispute resolution certainly did not "inspire general confidence" as we saw in Chapter 5 that it should. In fact, MIT never reviewed the notebooks. In June 1988, years after Margot O'Toole's initial complaint that some claims in the *Cell* paper were unwarranted, a congressional committee headed by Congressman Dingell subpoenaed the notebooks. That committee asked molecular biologists to look over the notebooks, but none were willing to get involved, including some molecular biologists who became highly critical of Baltimore and Imanishi-Kari.

Finally, the committee turned to a life scientist who was not a molecular biologist but who had a long record of interest and service in the ethics of research practice. Nelson Kiang, a noted researcher in otolaryngology, head of a laboratory at Massachusetts Eye and Ear Infirmary and a faculty member at Harvard, the Massachusetts General Hospital, and MIT, agreed to read the notebooks. What he saw was that some pages of notes looked markedly different from others. The appearance of those pages was so unusual that he insisted that the records be shown to David Baltimore (who was vigorously defending the research) to see if that would change Baltimore's view of the matter. It did not. The committee then forwarded those pages of notes to the Secret Service, because of that agency's expertise in inks.[16] [This event has been widely misrepresented as bringing in the Secret Service (rather than scientists) to investigate lab notebooks.]

The ink experts at the Secret Service found that the inks used to record data on the disputed pages were not in existence at the time of the publication of the *Cell* paper. This news finally precipitated the retraction of the *Cell* paper. It also led to a finding of misconduct against Imanishi-Kari by the NIH ORI, a finding that was overturned on appeal. Disagreement continues over whether the newer pages were Imanishi-Kari's honest attempts to organize a mess of notes before turning them over to the Secret Service, fabrication of data, or something else.[17] Discussion of this case continues to be highly polarized. For a brief and balanced account of what seems to be the final outcome see "Profile: Thereza Imanishi-Kari" in the November 1996 issue of *Scientific American.*[18]

What has stimulated the current concern about research is not a large number of instances of scientific misconduct but the observation that the research community's ability to develop and support ethical standards and expectations has not kept pace with the major growth and change in scientific research in recent decades and that a disturbing number of reputable researchers are negligent and unwilling to acknowledge their mistakes. By their actions, such researchers call into question the ability of the research community to govern itself.

An instance that alerted many people to the problem was a long delay in the publication of an article about negligence in publication. After researcher John Darsee was found to have committed multiple acts of misconduct, NIH researchers Walter Stewart and Ned Feder made a thorough review of Darsee's articles. They found that many reputable researchers who had coauthored articles with Darsee (and some of the journals that had published his papers) had failed to exercise appropriate oversight for the quality of published research. The publication in the prestigious journal *Nature* of Stewart and Feder's article documenting this negligence was delayed for years by threats of lawsuits against the journal from some of the researchers who were identified as having made these errors. In 1987, *Nature* finally published the article, "The Integrity of the Scientific Literature."[19] The delay eroded confidence that the research community could discuss or remedy problems in its own ranks.[20]

In the wake of such experiences the research community has reaffirmed a strong standard of responsibility and accountability for authors. Both the first (1989) and second (1995) editions of *On Being A Scientist* state that:

> Unless responsibility is apportioned explicitly in...the paper, the authors whose names appear on a paper must be willing to share responsibility for all of it.

Thus, if one is not able to vouch for an entire article, it is prudent to spell out the nature of one's contribution – to say, for example, that "J. did the statistical analysis of the data for this study."

Some journals have yet to allow implementation of this standard. They are often worried about using up printing space for statements about what each author has contributed to a work. Compliance with the standard would support responsible behavior on the part of researchers.

AN OFFER OF AUTHORSHIP

A colleague from another country asks you for help in preparing an English version of his article for a journal published in English. After you offer comments that are mostly editorial in nature, this colleague insists that you should be a coauthor. You do not think your contribution warrants coauthorship. It does not seem to you that your colleague will gain any reflected glory by having your name on the article since you are a junior member of the faculty, although you do not know much about how these things look in the colleague's home country. You do not want to insult your colleague.

What can and should you do?

Having one's contribution to the research mentioned in an acknowledgment, or having one's previously published or presented work cited in a paper, does not entail accountability for the publication as authorship does. Therefore, it is not necessary to get researchers' permission to cite their work or acknowledge their contribution.

Because of the responsibility that goes with authorship, one can harm as well as benefit people by designating them as coauthors. Therefore, all authors need to consent to being authors and each should approve the final version of the paper. Some journals, such as those of the American Chemical Society, now require the author who submits an article to a journal for publication to attest that all authors have seen and approved the paper. Some journals even require that all authors sign the submission form saying that they approve the paper. I have been surprised to see how often researchers forget this or add names to their papers without the knowledge or permission of the coauthors. Surprise authorship seems especially common when a researcher, in a misguided spirit of generosity, includes junior colleagues or students as authors. One member of an engineering faculty said he found himself listed as *primary* author on an article he had never seen and regarded as the work of a respected colleague, who was presumably the coauthor who submitted the paper.

Although faculty members are surprisingly lax about obtaining the consent of junior colleagues before adding them to an author list, some faculty members have made clear that they would take strong disciplinary action against any student who presumed to put their (the faculty member's) names on papers without first obtaining their permission. Some laboratory heads have the same proprietary feeling about the names of their laboratories. They object to a student submitting even a single-author publication without the lab head's knowledge, assuming the paper in some way identifies the lab at which the work has been done.

Discussing credit before submitting a work for publication is prudent as well as ethically responsible. A set of questions about a research supervisor's policies and practices for assigning credit used by graduate students to discuss credit with potential supervisors may be found in the WWW Ethics Center for Engineering and Science.

A rule requiring that all authors see and approve a manuscript before submission does not eliminate all difficulties, as the following scenario illustrates.

THE UNRESPONSIVE AUTHOR

You are a faculty member supervising a group of graduate students. Each student's thesis research has been supported through graduate school from the same funding source and each one's thesis supplies a piece of the total picture of the phenomenon you have funds to investigate. One of your students has accepted a position with a very large starting salary from an investment banking firm, to begin immediately after graduation. She accepted the position and completed the thesis research but barely completed the formalities to graduate. You now need to get her thesis research published to fill in that piece of the total research picture. You have repeatedly tried to contact her but have received no response.

What can and should you do now?

Along with proper inclusion of authors and acknowledgment of research contributions, proper citation of sources is crucial to the fair assignment of credit. It is also important for locating a research contribution in relation to the rest of the literature. The ethical guidelines of several professional societies state clearly that not only are authors expected to cite those publications that have been influential for the reported work but that they are "obligated to perform a literature search to find, and then cite, the original publications that describe closely related work."[21]

ETHICAL GUIDELINES FOR THE PUBLICATION OF RESEARCH

Since 1985, realization of the need for clear norms and expectations for responsible behavior in publication has led an increasing number of professional and scientific societies to make explicit rules and ethical expectations for editors and reviewers, as well as authors.

The Council of Biology Editors has published an extensive discussion of the ethics of publication in medicine and biology, where the nature of research, especially large clinical studies, raises distinctive issues.[22] The American Chemical Society, one of the most active organizations in the area of professional ethics, and a society that publishes several journals, took the lead in devising a thoughtful, detailed set of guidelines, which has since become a model for other professional societies, especially those in the physical sciences.[23] Journals generally publish their instructions and expectations for papers at least once a year in the journals. Increasingly, these statements go beyond formatting instructions and include ethically significant items. Some journals run by societies send the guidelines to any author who submits a manuscript. Only a few are as detailed as the ACS guidelines.[www] Twice, the ACS has updated them to address such new topics as publishing outside the technical literature.

These guidelines represent extended reflection on the ethically significant problems that arise for authors, reviewers, and editors of technical publications, but judgment is required to apply the more general guidelines appropriately. Those rules that are highly specific and require little judgment in their application are also those most likely to vary from one journal to another. Rules about the circumstances under which a reviewer of a manuscript may show a manuscript to another person also vary.

The specific rules usually have a common ethical basis, such as respect for the confidential nature of unpublished work and the authors' proprietary interests in it, but reasonable and responsible people may differ on the specifics. Very specific rules represent expectations for publishing in the journal that issued the rule, and by submitting to that journal one takes on an obligation, rather like the obligation of a promise, to abide by their rules in the submission in question. Both the general ethical considerations that underlie specific rules and the rules and obligations that are more give guidance about responsible behavior, but they leave specific action to the discretion of the author.

Item #11 of the "Ethical Obligations of Authors" section of the ACS guidelines, which states appropriate criteria for inclusion as an author, illustrates many of the points about the ethical basis and function of provisions of ethical guidelines on publication.

> The co-authors of a paper should be all those persons who have made significant scientific contributions to the work reported and who share responsibility and accountability for the results. Other contributions should be indicated in a footnote or an "Acknowledgments" section. An administrative relationship to the investigation does not of itself qualify a person for co-authorship (but occasionally it may be appropriate to acknowledge major administrative assistance). Deceased persons who meet the criterion for inclusion as co-authors should be so included, with a footnote reporting date of death. *No fictitious name should be listed as an author or co-author.* The author who submits a manuscript for publication accepts the responsibility of having included as co-authors all persons appropriate and none inappropriate. The submitting author should have sent each living co-author a draft copy of the manuscript and have obtained the co-author's assent to co-authorship of it. (Italics added)

These guidelines say that the submitting author is responsible for ensuring that all have seen the final version, but some other journals say that all authors have to sign a statement saying they have seen and agree with the final version of a submitted manuscript. The importance of obtaining the consent of authors to being listed as authors is generally recognized by journals, and, as we saw, stems for the responsibility and accountability that goes with authorship. This is also the basis for the prohibition against the use of fictitious names in publication of research.

To publish pseudonymously in one of the ACS journals would be to fail to honor an explicit understanding about the conditions for publishing in that particular journal. Failure to live up to conditions agreed upon in particular circumstances is prima facie wrong. The common use of "pen names" in fiction writing is widely approved, however, and suggests that there may be nothing wrong with fictional authorship per se. Consider a situation in which a group of researchers band together to publish under a pseudonym.

A novel, *The Bourbaki Gambit*, describes an effort at joint publication that serves as an opportunity for the authoring group to rise above the desire for individual fame. This is the second of what will be four novels about the moral life of research scientists by Carl Djerassi, an internationally known chemist and recipient of both the National Medal of Science and the National Medal of Technology. The title of this acclaimed novel refers to a group of mathematicians who all published under the pseudonym "Bourbaki." Djerassi's novel, like the original Bourbaki endeavor, involves a close collaboration and not merely a shared pseudonym and raises interesting questions about the virtues and vices fostered by a life in research. In the foreword to his book,

> **Nicolas Bourbaki, Fictitious Author**
> Since the 1930s a group of mathematicians has been publishing a treatise on the whole of mathematics using the pseudonym Nicolas Bourbaki. Over two dozen volumes had appeared by the 1960s. The exact membership of the group was kept secret, although certain well-known mathematicians are generally believed to have been a part of the effort. Younger mathematicians joined the effort over time.
>
> Bourbaki's work is marked by an axiomatic method, strict logical arrangement, generality, and a carefully chosen terminology. It has proved to be extremely influential on subsequent mathematical research.

Djerassi describes his theme as "the twin issues of collegiality and the desire for fame, pointing to the tension between the collaborative effort at the heart of modern science and the desire for individual recognition in the hearts of most scientists."[24] Responsible means of coping with that tension requires more than merely following rules. If accountability were maintained in some way, the moral acceptability of transcending that tension with a Bourbaki-type collaboration and pseudonymous publication in science or engineering journals that did not expressly forbid it might be seen at worst strange and annoying, rather than unethical.

If such collaborations proved common and fruitful, the specific rules blocking them would undoubtedly change.

To learn the most from a set of ethical guidelines, one must understand what is fundamental in them. Consider first why the guidelines mention the issues that they do. As we have seen, codes and guidelines are developed to guide people through situations that commonly occur in a particular context. The guidance they give and the limits they set reflect both the situations they anticipate and the values they mean to uphold. If one misunderstands the context or the problem situations anticipated, then one is liable to misinterpret the values the guidelines are intended to promote. At the same time, if the drafters of the guidelines misunderstand the context or (more plausibly) the problem situations that do arise, then the guidelines may fail to promote the intended values or may even undermine them. (As we saw with codes of ethics of professional societies, the fact that something is an ethical guideline does not put it beyond ethical criticism or even guarantee that it concerns a matter of ethics.)

Two sets of guidelines that differ may reflect a difference in priority accorded certain values as well as in the specific actions they expect to implement those values. Disciplinary and field differences also influence the situations the guidelines cover. Such differences influence how many researchers are needed to conduct a typical experiment, for instance, or the role of lab heads or the presence of post-docs.

Consider, for example, the standards of professional responsibility required of reviewers for journals whose editors are full-time editors rather than researchers. The majority of technical journals are edited by researchers who are able to independently assess most manuscripts submitted to them. Reviewers only make recommendations to these editors. The editors of some prestigious physics journals are not researchers, however. Those editors are much more dependent on reviewers' reports. To be fair, reviewers need to be scrupulous in confessing their biases and give the editors a clear picture of the strengths and weaknesses of an article. This is even more important if the journal uses only one reviewer per article.

We saw in Chapter 2 that in engineering ethical codes and guidelines, and especially in some of the decisions by the NSPE's Board of Ethical Review, treatment of one's fellow engineers and concern for the dignity of the profession receive strong emphasis. Similarly, in ethical guidelines for the publication of research, we find a prominent concern to eliminate practices that waste the time and energy of reviewers and editors. Indeed, the second item under the "Ethical Obligations of Authors" section of the ACS guidelines says that:

> An author should recognize that journal space is a precious resource created at considerable cost. An author therefore has an obligation to use it wisely and economically.

Unwise use of the publishing privilege wastes the time of reviewers and editors. It is a minor rather than serious offense, but, like littering, its cumulative effect diminishes a common resource.

Additional evidence of the concern about publication resources is found in Items 7 and 8 of the list of obligations of authors: Potential authors are admonished not to submit the same research to two journals simultaneously (#8), and to inform an editor of related manuscripts under editorial consideration or in press elsewhere (#7). The purpose of these two ethical guidelines is to prevent authors from wasting the time of one journal's editor and reviewers, or using the resources of two journals to make only one contribution to the literature. Although failure to follow these two ethical guidelines neither compromises research integrity nor jeopardizes the fair assignment of credit for work, it does violate the understanding underlying submission of a manuscript. Therefore, doing so is deception and the rule against it is an ethical stricture and not a mere matter of "etiquette."

A violation of the rule against submitting the same research to two journals simultaneously, and that requiring authors to inform an editor of related manuscripts under editorial consideration or in press elsewhere, occurred recently. In ignorance of the other article, the editors of the *New England Journal of Medicine* and *the Journal of the American Medical Association* each published articles reporting the same research within a matter of a few weeks of each other. Needless to say, the editors of these two prestigious journals were displeased, and the authors obtained increased exposure for their article at the cost of a dubious reputation for themselves.

The more familiar and clearly ethical concerns about such matters as fair crediting of other sources also receive extensive attention, along with guidance about the different criteria used for reports of research and for review articles:

> An author should cite those publications that have been influential in determining the nature of the reported work and that will guide the reader quickly to the earlier work that is essential for understanding the present

investigation. Except in a review, citation of work that will not be referred to in the reported research should be minimized. An author is obligated to perform a literature search to find, and then cite, the original publications that describe closely related work. For critical materials used in the work, proper citation to sources should also be made when these were supplied by a nonauthor.[25]

The safety hazards in chemical research receive recognition in the reminder to authors to identify any unusual hazards in the conduct of their research. (Although any researcher should disclose such hazards, they would be such a rarity in some areas of research as to go unmentioned in guidelines of journals in those disciplines.)

The ACS guidelines enjoin authors to avoid fragmenting their research. The fragmentation of research has increased as investigators have perceived that the *number* of their publications was of overriding significance. This perception arose as grant reviewers and even promotion and tenure committees sought to simplify their evaluation task by counting the number of their publications. Indeed, at some universities, junior faculty, especially those in medical fields, have been told that a criterion for tenure is a certain number of publications (and continued grant support). This practice has given rise to the tongue-in-cheek expression "lpu" for "least publishable unit" as a measure of research productivity. To encourage quality rather than quantity, some universities now limit the number of publications that a candidate can offer for evaluation when being considered for promotion or tenure.[26]

More obvious than an authors' responsibility to provide complete research statements is the responsibility to correct errors.[27] The guidelines also mention a journal editor's responsibility to see that published errors are corrected either by the author of the original article or by the person who has discovered the error. (Authors, of course, also have a responsibility for seeing that errors in their work are also corrected. As we saw in the introduction on concepts, moral responsibility, unlike official responsibility, is not exclusionary.)

Reviewers and editors, as well as authors, have ethical responsibilities in the conduct of their work. Editors as well as research institutions and government agencies must deal with allegations of misconduct on the part of authors as reviewers, and since the mid-1980s, editors as well as research institutions have been learning how better to respond.

The case of Vijay Soman's flagrant plagiarism of a manuscript he reviewed illustrates the institutional mishandling of misconduct cases. Philip Felig, Soman's senior colleague and former advisor at Yale, asked

Soman to help review a paper submitted for publication by two NIH researchers, Helena Wachslicht-Rodbard and Jesse Roth. The work was similar to a project Felig had begun but not completed some years earlier. Soman recommended that the paper be rejected, and he himself then took up the project and completed it the next year. He submitted a paper citing Felig as coauthor.

Fortuitously, the journal sent the paper to Roth to referee who showed the paper to Wachslicht-Rodbard. She recognized that the paper plagiarized sections of their rejected paper and notified the journal editor. Both Roth, who knew Felig well, and the journal editor called Felig. Felig assured them that the work had been done independently but agreed to modify some of the identical passages and to delay publication so that Wachslicht-Rodbard's article would appear first. Roth agreed to this remedy, but Wachslicht-Rodbard did not and wrote to the dean of the Yale Medical School charging plagiarism and data fabrication. Both Felig and the dean treated the actions as minor transgressions. When Wachslicht-Rodbard threatened a public denunciation of Felig and Soman, Roth and Felig called in an auditor who stalled for eight months, during which time the paper by Soman and Felig appeared, despite Felig's earlier offer to delay. Wachslicht-Rodbard then demanded that Roth take immediate action, so he arranged for another auditor to inspect Soman's work. This auditor found serious discrepancies. In the end, Felig withdrew twelve papers of Soman's, including eight on which he was coauthor.[28]

This case was one that helped convince the research community that the mechanisms for investigating charges of misconduct needed reform. It should not, however, lead to the conclusion that only reviewers (or "referees") whose behavior is as extreme as Soman's misuse manuscripts they review. The more common abuse by reviewers is to take unfair advantage of what they learn from reviewing a manuscript or a grant proposal.

CONFLICT OF INTEREST AND POTENTIALLY CONFLICTING INTERESTS

THE REVIEWER WHO BECOMES INTERESTED[29]

The editor of a journal in your field asks you, a recent Ph.D., to review a manuscript that has been submitted to that journal. This manuscript provides a proof for a result in your area. You become intrigued by the topic and after a week or two come up with a shorter and better proof. You feel clear about your recommendation regarding the publishability of the result, but what, if anything, should you say about your new proof?

Two sorts of interests are commonly involved in conflict of interest: 1. interest in competitive research advantage and the career advancement that attends it and 2. financial interest. Both sorts may arise in the context of scientific publication. Conflict of interest in the review of manuscripts and grant proposals is a major concern in research ethics. The ACS Guidelines for reviewers and editors emphasize this point. The obligation for editors is somewhat more stringent: Editors are admonished to pass the editorial responsibility for a manuscript that is closely related to their own past or present research on to another person. Reviewers are advised to be sensitive to the appearance of a conflict of interest in such cases, and if in doubt, to return the paper advising the editor of the bias or conflict of interest or alternatively to furnish a signed review stating the reviewer's interest in the work, with the understanding that the editor may pass this signed review to the author. (Normally, the identity of reviewers is kept confidential.) The difference in guidelines for editors and reviewers stems from the difficulty in getting a knowledgeable review unless the reviewer also works in that area. The assumption in these guidelines is that the editor is also a researcher rather than a professional editor. In that case, the editor may recognize and control the bias of a reviewer. Because there is no one to compensate for the editor's bias, the rules bearing on control of bias and conflict of interest by editors are more stringent.

The ACS guidelines allow that if either an editor or a reviewer learns from a manuscript that some of their current work is likely to be unprofitable – say the reviewer is looking for a certain gene in one area of a chromosome and the author has found it elsewhere – the editor or reviewer may discontinue the work. This rule is reasonable, and it shows that those who wrote the guidelines thought through many of their implications and so did not issue the simplistic rule that reviewers and editors may make no use of anything they learn from an unpublished manuscript.

The guidelines do not, however, simply specify ethically acceptable and unacceptable behavior. Consider the following scenario:

RESPONSIBILITIES OF A REVIEWER

Your thesis supervisor asks you to help in reviewing a journal article that has been sent to her by the editor of a prominent journal, because the topic of the paper is related to some of your own work. As you study it, you find the study to be quite significant and you realize that one of its implications is that the research being pursued by a good friend of yours leads up a blind alley.

What should you do and how should you go about it?

Will it make a difference to what you do or the way you go about it, if the student is working with the same faculty member?

The issue in this scenario is of treating the manuscript as a confidential document, so that it can be shown or, presumably, described only to those whose opinion is needed to help in its evaluation. Even then, one should at least inform the journal editor. (As we noted earlier, journals vary on the specific rule. Some go further than the ACS Guidelines and require a reviewer not merely to inform but also receive the *prior permission* of the editor to show the manuscript to others.) Unless the author has given permission to say more, the editor may not share information about the manuscript, other than to list authors and titles of articles accepted for publication.

Tensions between interests or responsibilities do not necessarily create conflicts of interest, situations that one is morally required to avoid or disclose, but are tensions with which a responsible practitioner is expected to cope when making professional judgments. As we saw in Chapter 2, in a conflict of interest situation, it is not sufficient to be knowledgeable and conscientious in exercising professional judgment. A conflict of interest situation arises when one is expected to avoid or change to eliminate the possibility of even unconscious bias, or if that is not possible, to fully disclose. Determination of just what constitutes a conflict of interest situation changes over time, sometimes because of a recognition that too many practitioners or office holders have failed to manage some tension well. Some potentially conflicting interests, obligations, or responsibilities could not be treated as conflicts of interest since they are permanent features of professional practice. It would, therefore, be pointless to disclose them (everyone knows about them), and the practitioner cannot eliminate them. For example, academic researchers have many responsibilities and obligations: for the advance knowledge in their field, to their research sponsor, to take a fair share of the unpaid work (such as reviewing journal submissions and grant proposals) in their field, to be fair to other researchers in that process, for the support of family members who depend upon them for their livelihood, etc. These may come into conflict with each other or with their own interest in advancing their careers. Usually, none of these responsibilities can be divested.

Financial conflicts of interest have received much attention in recent years. This type of situation arises when a person in a position of trust is required to exercise judgment on behalf of others but also has specifically financial interests of the sort that might interfere with the exercise of judgment in that position of trust. Indeed, policies requiring disclosure of financial interests that might conflict with judgment as a researcher or as a public official are very commonly called "conflict of interest policies," as though financial conflict of interest were the definitive or perhaps the only conflict of interest. The ACS guidelines holds that authors are bound to disclose any financial conflict of interest. They say:

> The authors should reveal to the editor any potential conflict of interest, e.g., a consulting or financial interest in a company, that might be affected by publication of the results contained in a manuscript.

1993 legislation reauthorizing the operations of the National Institutes of Health (NIH) required new federal regulations to minimize conflicts of interest among scientists supported by the NIH, including those receiving NIH grants. The regulations have now been issued and require disclosure of financial interests of more than $10,000. This legislation and the regulations resolve a five-year controversy in which public support for regulations to minimize financial conflicts of interest, especially in the evaluation of a medical product or treatment, was met by resistance at several major medical schools. Health policy analyst, Diana M. Zuckerman, summarizes the issues in the following way:

> It is a simple fact of life that scientists who study particular drugs could be influenced by having financial arrangements with the companies that make those drugs. Many ways exist to analyze data, and if the results are not dramatic one way or the other, a scientist could be motivated to find a significant result where none really exists, to omit some potentially relevant information, or even to unintentionally skew the findings.

That kind of bias is probably inevitable and thus has to be considered acceptable, for example, when scientists working for a particular company are in the early stages of product development. However, such bias becomes unacceptable when the public, not the company, pays for the research, especially when the public could be put at risk as a result.[30]

Areas of conflicts of interest may exist among the conflicting responsibilities of academic researchers mentioned earlier, and these are important for their graduate and sometimes even undergraduate students. So much so that of the nine scenarios provided for student–faculty discussion in *On Being a Scientist*, two concern the implications for graduate students of the consulting agreements or industrial sponsorship received by their research supervisors. One of these, the one that concerns the consulting relationships of faculty, is even titled "A Conflict of Interest."[31] Industry sponsorship of research, for decades common in engineering, is now increasingly common in many areas of scientific research.

Here are another series of scenarios that set out issues somewhat more starkly:

WORKING FOR THE COMPANY – 1

You are a graduate student in the doctoral program of Major Engineering Department. You have made excellent progress thus far and at the end of your second year you are ready to begin your thesis. You have been working very hard and have done well in all your subjects despite carrying an excess course load (as your fellowship permitted you to). You know you will need to keep up this pace in order to finish in three years. That time limit is dictated by (choose one):

- stringent family obligations, or
- the provisions of your visa and fellowship, or
- your active duty status in the armed services.

You need a topic that you will be able to complete in a little more than a year. After speaking to several potential thesis supervisors, one of them, Professor A., mentions two topics that sound feasible, if not very exciting. Both of these, indeed all of the topics mentioned by Professor A., are topics that are directly related to the research program of the company Professor A. just founded.

What questions should you ask, and of whom, before deciding?

WORKING FOR THE COMPANY – 2

Suppose that you know two other thesis students currently working with this professor and they tell you that the Professor regularly asks them to drop their normal education and research activities to run down to his company and give a presentation, impart a little know-how, or the like.
Now what do you do?
Where can you go for advice?
What might the other two students do?   Where might they go for advice?

Some universities or university departments have explicit provisions for maintaining quality control on education, preventing problems, and curtailing abuses. These include: limiting the amount of outside consulting or work on one's own company that full-time faculty are permitted to do to one day a week, defining it as a conflict of interest for a faculty member to hire his own thesis students in his consulting, and forbidding the practice. The goal is to keep thesis supervision free of influence of faculty member needs from assistants to fulfill his consulting agreements. Other provisions will be discussed in the next section.

CREDIT ISSUES AMONG FACULTY, POST-DOCS, AND
GRADUATE STUDENTS

The question of apportioning credit between junior and senior researchers is coming to be recognized as an important and delicate issue, so important as to be given its own section in the first (1989) edition of *On Being a Scientist*, and it receives much discussion in the most recent (1994) edition. The unequal power between junior and senior researchers creates special moral hazards, especially because power differentials are likely to be underestimated by the more powerful party. Furthermore, when the junior person is the student or trainee of the senior person, the responsibility of the senior person to educate the junior person in her rights and responsibilities may directly conflict with the senior person's self-interests or the perceived interests of the laboratory or research group that the senior person heads. The relative absence of general discussion of research ethics in recent decades has led to the inarticulateness on the subject on the part of many fair-minded and decent researchers. This inarticulateness has stood in the way of the development of group norms and informal sanctions. None of the authoritative statements on the fair treatment of supervisees give specific standards. Despite the clarity of guidance that the latest (1995) edition of *On Being a Scientist* gives about other matters, on the subject of the apportionment of credit between junior and senior authors it says only that:

> Senior scientists are...expected to give junior researchers credit where credit is warranted. In such cases, junior researchers may be listed as co-authors or even senior authors, depending on the work, traditions within the field, and arrangements within the team.[32]

True as this statement may be, it is of little help to graduate students in assessing their own experience with their supervisors. Graduate students and post-doctoral fellows are dependent on their research supervisors for advancement in their early careers. Students and fellows are especially in need of low-risk ways of understanding and assessing matters of research ethics, including their own treatment. Educational efforts that build the competence of the faculty to talk about their practices are needed, and such activities are worth looking for in a graduate department.

Another feature of a well-run graduate department that recognizes the responsibility of a department for the development of its graduate students is the presence of guidelines and requirements for faculty with regard to such matters as the formation of thesis committees and the frequency with which they are required to meet and report on a student's progress. The oversight that a thesis committee is expected to exercise will not occur if the committee does not meet. The committee may not meet even yearly if meetings are left to the student to try to arrange in the face of busy faculty schedules. Another significant matter is the support offered to a student who wishes to terminate a relationship with one research supervisor and find another. Another positive sign is that a department makes an effort to assign graduate students a departmental advisor different from the student's research supervisor, so that it is relatively easy for a student to get a "second opinion" from some faculty member who has an ongoing relationship with the student. In some departments the graduate officer or faculty member in charge of graduate studies is a source of unbiased advice about handling difficult situations. There also may be a dean of graduate studies at the university or a university ombudsperson who provides such help and advice.

THE REVIEW OF GRANT PROPOSALS

WHAT BECAME OF MY RESEARCH PROPOSAL?

Several years ago, you proposed a research project to a funding agency. Although it was rated as strong, it was not rated highly enough to be funded. The proposed research had three parts. Most of the criticism was directed against two of the three parts.

You forgot about the proposal until you came across a journal article that is incredibly similar to the work proposed in the strongest segment of your proposal.

Is there any ambiguity in this situation?

If your ideas have been used, how serious is this misuse of the proposal?

What can or should you do?

How much difference does it make that the writers have or have not appropriated whole sentences from your proposal?

Journal reviewers and reviewers of grant proposals submitted to funding agencies share many responsibilities. Both see others' ideas before they have been published, although, admittedly, at different stages of their development. The abuses I hear of most often are reviewers' plagiarism of ideas, and occasionally of words, formulas, and diagrams from grant proposals. To make the theft effective, the reviewer assigns a proposal so low a score that it is not funded. Plagiarism by a grant reviewer is thus a double offense: plagiarism and unfair evaluation.

The most documented instances of misconduct on a grant proposal have involved plagiarism rather than fabrication (as in the scenario) or falsification. This higher rate of reported plagiarism may be due to a greater likelihood that plagiarism will be detected, that it will be reported (grant reviewers who find their own work plagiarized report it), or that plagiarism is more common than other misconduct.

The high incidence of misrepresentation in grant proposals, as contrasted with publications or papers delivered to professional groups, is attributed to perpetrators' mistaken belief that dishonesty is an excusable part of "grantsmanship," that is, the effective marketing of proposals for research. Funding agencies and many universities, however, regard deception in grant proposals as a serious matter. Universities have fired perpetrators, and granting agencies have debarred them from grant competition.

CONCLUSION

Competition must be fair to provide the context of trust and cooperation on which advances in research depend. Standards of fairness are embodied in practices for crediting research contributions. The most serious violation of these, plagiarism, counts as research misconduct, but many subtler considerations of fairness also exist. As with other practices examined in Chapter 6, these standards have evolved over time and some of the specific provisions vary from one field to the next. Even where the specific requirements are not uniformly recognized in all science and engineering fields, guidelines on ethical publication practices, such as the one issued by the ACS, provide very useful aid to identifying the ethically significant factors and circumstances one should consider to be fair and responsible in publishing one's research.

Conflict between faculty members and either post-docs or graduate students over credit and control of research is an important issue. It needs discussion, since misunderstandings are difficult to clarify if the research supervisors are the students' or post-docs' only source of information, and since misunderstandings as well as actual violations of standards can do important damage to the transmission of ethical norms in research practice.

Although grant proposals are not, strictly speaking, publications, many of the same factors are relevant to ethical practice in writing and reviewing grant proposals.

NOTES

1. This case is loosely based on NSPE Case No. 85-1.[www]
2. Based on a scenario by Giovanni Flammia, computer science graduate student, MIT.[www]
3. This Case, "Credit for Engineering Work," and the Board's Discussion appear in *Opinions of the Board of Ethical Review*, Volume VII. Alexandria, VA: National Society of Professional Engineers. 1994, pp. 81–82. It is on the Engineering Cases Section of the WWW Ethics Center for Engineering and Science.
4. *On Being a Scientist*, p. 9.
5. Potter, Elizabeth. 1994. "Locke's Epistemology and Women's Struggles." In *Modern Engendering*, edited by Bat Ami Bar On. New York: SUNY Press.
6. Kuhn, Thomas S. 1997. *The Essential Tension*. Chicago: University of Chicago Press.

7. C. K. Gunsalus commenting on her experience handling allegations of misconduct at research institutions, meeting of the AAAS Committee on Scientific Freedom and Responsibility, September 1993.
8. Kaufman, Ron. 1992. "After 5 Years, Heated Controversy Persists In Science Copyright Case." *The Scientist*. September 14: 1, 4, 5, 10.
9. National Academy of Sciences' Convocation on Scientific Conduct, June 1994. Public discussion.
10. "Senior author" is also sometimes used as a synonym for "primary author." For example, Carl Djerassi so uses it in Chapter 8 of *Cantor's Dilemma*.
11. Djerassi, Carl. 1989. *Cantor's Dilemma*. New York: Penguin Books, USA. See especially Chapters 8 and 9.
12. Djerassi, Carl. 1994. *The Bourbaki Gambit*. Athens: University of Georgia Press.
13. Daniel Kevles reported on media exaggerations at a symposium, *Government, the Media, and Scientific Misconduct: The David Baltimore Case in American Political Culture* held at MIT, October 28, 1996. His comments will be generally available in his forthcoming book on the case.
14. See Weiss, Philip. 1989. "Conduct Unbecoming?" *The New York Times Magazine*, Oct. 29: pp. 40–41, 68; and Wheeler, David L. 1992 "U.S. Attorney: Leave 'Baltimore Case' to the Scientists." *The Chronicle of Higher Education*. July 22, 1992, p. A7.
15. Having witnessed several aspects of this case at very close range, my own view is that if MIT had obtained the laboratory records promptly, the matter could have been completely resolved on the spot without any charges of misconduct being leveled. MIT was not unique in its omission to do so, however. In 1986 many universities had not yet learned to promptly obtain the lab records. For discussion of some current procedures and guidelines and their role in preventing misconduct see Shore, Eleanor G. 1995. "Effectiveness of Research Guidelines in Prevention of Scientific Misconduct." *Science and Engineering Ethics*, 1(4): (October).
16. Personal communication, Nelson Kiang. I subsequently taught both an undergraduate course in engineering ethics and a graduate course in research ethics with Nelson Kiang and had numerous opportunities to discuss the events with him.
17. These events and the variety of interpretations to which they are subjected to are discussed in Sarasohn, Judy. 1993. *Science on Trial*. New York: St. Martin's Press, Chapter 7.
18. Beardsley, Tim. 1996. "Profile: Thereza Imanishi-Kari." *Scientific American*, 275(5) (November): 50–52.
19. Stewart, Walter and Ned. Feder. 1987. "The Integrity of the Scientific Literature." *Nature*, 325(January 15): 207–214.
20. The embargo of Stewart and Feder's article was first brought to my attention by Robert W. Mann, a world famous biomedical engineer. In my experience, academic engineers are significantly more willing to talk about ethical standards and what can be done about violations of them than are researchers in other disciplines. This may stem from the engineering profession's history of concern with professional ethics, but I also note that this willingness increases with the professional stature of the engineer.

21. American Chemical Society. 1994. "Ethical Guidelines to Publication of Chemical Research." *Accounts of Chemical Research* 27(6):179–181. This is the most recent revision; the first version was written in 1985. The American Geophysical Union and the Optical Society of America have issued their own guidelines based on those of the American Chemical Society.

22. Bailar, John C., Angell, Marcia, Boots, Sharon, Heumann, Karl, Miller, Melanie, Myers, Evelyn, Palmer, Nancy, Weinhouse, Sidney, and Woolf, Patricia. 1990. *Ethics and Policy in Scientific Publication.* [The Committee on Editorial Policy, Council of Biology Editors (CBE).] Bethesda, MD: Council of Biology Editors, Inc.

23. American Chemical Society. 1994. "Ethical Guidelines to Publication of Chemical Research." *Acc. Chemical Research,* 27(6):179–181.

24. Djerassi, Carl. 1994. *The Bourbaki Gambit.* Athens: University of Georgia Press.

25. Item 4 under "Ethical Obligations of Authors" in American Chemical Society. 1994. "Ethical Guidelines to Publication of Chemical Research." *Accounts of Chemical Research,* 27(6):180.

26. Culliton, Barbara J. 1988. "Harvard Tackles the Rush to Publication." *Science,* 241 (July 29, 1988): 525.

27. See, for example, *Guidelines of the Optical Society of America Concerning Ethical Practices in the Publication of Research,* available from the Optical Society of America.

28. See Mazur, 1989, and Broad and Wade, 1982, pp. 75–77.

29. Adapted from a case by Michael Lavine, Statistics Department, Duke University.

30. "Conflict of Interest and Science." *The Chronicle of Higher Education,* October 13, 1993, p. B1.

31. *On Being a Scientist,* pp. 8, 9.

32. *On Being a Scientist,* p. 14.

10

CREDIT AND INTELLECTUAL PROPERTY IN ENGINEERING PRACTICE

PATENTS AND TRADE SECRETS

INTELLECTUAL PROPERTY OF ENGINEERS IN PRIVATE PRACTICE www

Roy, an engineer, submits a proposal for a city project to the county council. The proposal included technical information and data that the council requests. A staff member of the council makes Roy's proposal available to Thornton, another engineer. Thornton uses Roy's proposal to develop another proposal for a somewhat different project and submitted it to the council. The parties dispute the amount of Roy's information that Thornton used.

Is Thornton guilty of plagiarism? Does it make a difference, ethically speaking, if the amount of information used was large or small?

What are the city council's responsibilities in handling Roy's proposal?

THE USE OF WORK FROM AN UNPAID CONSULTATION www2

A state agency considers designing a facility that requires special expertise in the field of solar energy. They learn from a federal agency that the Moreau firm had previously developed a plan for a similar facility for that agency, and so they contact the Moreau firm. The Moreau firm submits preliminary data to the state agency, who in turn includes that information in a proposal to a private foundation to secure additional funds for the project. The state agency holds many informal discussions with Moreau's firm and so leads that firm to believe that, if the project is approved, it will be awarded the contract.

295

Several months later, the state agency tells Moreau's firm that the public and private funding it received will not be sufficient to fund the full scope of the facility. The firm is then asked to evaluate the possibility of a more limited facility. Believing that it will be awarded the design contract, the Moreau firm investigates the possibility for a more limited project at its own expense of several thousand dollars and submits a revised proposal to the state agency.

Subsequently, the chief state engineer informs Moreau's firm that he had turned over all of its data to the Barron firm and is conducting initial negotiations with them. The chief says that if these negotiations fall through, it will contact Moreau's firm to negotiate the project. All the while, the Barron firm had been aware of the involvement of the Moreau firm in the project, but it has not contacted them to discuss the project or to get its earlier submissions to the state agency. The Moreau firm protests to the state agency, accusing the Barron firm of violating the Code of Ethics.

Does Moreau's firm have an implicit contract with the state agency?

Is their provision of unpaid services an improper gift to the state agency?

How would you evaluate the conduct of the firms? Of the state agency?

Is there other information that would be morally relevant in evaluating the situation?

IS IT WRONG TO COPY A VENDOR ID? [www3]

SCSI, an industry standard system for connecting devices (like disks) to computers, provides a vendor ID protocol by which the computer can identify the supplier (and model) of every attached disk.

First Company makes file servers consisting of a processor and disks. Disks sold by First identify First in their vendor ID. Disks from other manufacturers can be connected to First's file servers; however, the file server software can perform certain maintenance functions (notably prefailure warnings based on performance monitoring) only on disks made by First Company.

Competitor Company decides to compete with First by supplying cheaper disks for First's file server. They quickly discover that although their disks work on First's file servers, their disks are at a disadvantage because they lack the prefailure warning feature of First's disks. The CEO of Competitor, therefore, directs the engineer in charge of the disk product to "find a solution to this problem." The engineer uses reverse engineering and discovers that by making the vendor ID on their cheaper disks match that of First's disks, the First file servers will treat Competitor disks as First disks. Competitor incorporates this change into its product and advertises the disks as "100% First-compatible."

Representatives of First charge Competitor with forgery; they are convinced that, whether or not Competitor's practice is illegal, it clearly violates industry-wide ethics.

Competitor justifies its action on the grounds that the favored treatment of First's disks by First's servers is unfair and monopolistic. Moreover, they argue that using First's vendor ID isn't forgery, since it doesn't mislead people: Competitor's disks are clearly labeled as coming from Competitor. Their action at most 'misleads' First's software. If this action is not forgery, what is it? What, if any, legal or ethical rights are infringed by copying the vendor ID of the Competitor disks?

Design work and other innovative technical contributions are recognized in a variety of ways in a variety of settings. Credit for a design may be reflected by naming the device for an individual (for example, the Jarvik heart) or to a group or corporation (for example, an NCS knee or "Ford's new utility vehicle)." Even when a device is named for a person, that individual need not be the designer. For example, many medical devices that carry the names of individuals are named for the physician who collaborated on it or who perhaps was the first clinician to use the device.

The research that goes into a new product and the testing undertaken to identify the causes of some defects that appear in production are rarely published in journals or books. Research and testing in industry are commonly credited in the same way as design and manufacturing work, that is, by promotions, raises, and the like, rather than by ownership or association of one's name with a design or device. The products of work done for an industrial employer are generally recognized as the employer's property. The NSPE stipulates that an "engineer's designs, data, records, and notes referring exclusively to an employer's work are the employer's property." In contrast, the research done by a university laboratory working under a grant or a contract is presumed to be something that the researchers may publish (perhaps after an agreed-on delay to give an industrial sponsor a head start in using the results.)

The proprietary rights embodied in patents and copyrights work differently from crediting mechanisms that have no property implications. The patent arrangements that attend industrial sponsorship of university research are more complex than authorship credit. Recall the adaptation of NSPE case 92-7, in which the head of a company offered to fund two faculty members of a university's chemistry department for research on removing poisonous heavy metals from waste streams. Under that agreement the university agreed to give the company exclusive use of any resulting technology developed for water and waste water treatment. The company left the faculty members free to exploit applications of the technology other than the treatment of water and waste water. Presumably any patent on the technology would belong to the university or its faculty members, with the company having exclusive license to develop or apply the technology for water treatment.

Ethics codes and guidelines of engineering professional societies may provide further guidance. The section of the NSPE's code of ethics addresses much more than proprietary interests in discussing credit. It outlines the following as professional obligations:

10. Engineers shall give credit for engineering work to those to whom credit is due, and will recognize the proprietary interests of others.
   a. Engineers shall, whenever possible, name the person or persons who may be individually responsible for designs, inventions, writings, or other accomplishments.
   b. Engineers using designs supplied by a client recognize that the designs remain the property of the client and may not be duplicated by the Engineer for others without express permission.
   c. Engineers, before undertaking work for others in connection with which the Engineer may make improvements, plans, designs, inventions, or other records which may justify copyrights or patents, should enter into a positive agreement regarding ownership.
   d. Engineers' designs, data, records, and notes referring exclusively to an employer's work are the employer's property.

As we saw in the introduction, a patent is a legal right granted by the government to use, or at least to bar others from using, one's invention. In the United States this right lasts for seventeen years for useful devices and fourteen years for designs. This right may be assigned to others, so the owner of the patent or copyright is not necessarily the inventor or author, although the original patent document records the name of the inventors (the order of the listing is not significant). Because obtaining patents and defending them are costly, inventors sometimes prefer to assign them to others and receive alternative forms of compensation for their work.

Once one has decided to file a patent, questions of priority enter, much as they do in claims of scientific discovery. Submission for publication establishes the date and hence the priority claim for published work. In filing for a patent, the crucial date for claim of priority is not the date of the filing but the *date of conception of the idea,* although in the United States one must file within a year of any "public disclosure" of the design or plans.

Documentation of the date (and even time) when the idea was conceived is needed when competing applicants file for the same claims. Igor Paul, a professor of mechanical engineering who is familiar with the legal system, recommends two ways of proving when the ideas were conceived:

- Keep a permanent design notebook (bound, with sequentially numbered pages) documenting your work and ideas (dated and in ink), and have it periodically (or for a particular idea) dated and signed by your instructor.
- Document your idea with annotated sketches, explain it to a fellow student and/or instructor, have them sign the document and indicate that they have understood the idea, and then send the document to yourself by registered letter, which you save unopened. The postage date and time stamp on the letter documents the time you conceived the idea.

The rules for keeping a laboratory notebook on a research project are the same as those for keeping a design notebook. Your design notebook is your own possession, however, so you can take it, rather than only a copy of it, when you graduate.

In U.S. patent law, an application must be initiated within one calendar year of an idea's "public disclosure." The European and Japanese patent laws require patent application prior to public disclosure, but most countries honor patents taken out in other countries. Presentation in a class or elsewhere in a university community does not constitute public disclosure. If someone in the audience for, say, a class presentation, were to discuss the ideas outside the university, however, that discussion might constitute public disclosure. Since such disclosure may occur on the part of students, students need some understanding of the laws and conventions regarding intellectual property.[1]

Neither public disclosure nor filing for a patent are precise analogs of publication of research. Public disclosure makes work public and may enhance an inventor's reputation, but it is not a mechanism for establishing priority unless publication goes with public presentation. Filing for a patent establishes a *claim* to ownership, but, unlike the date of a publication, the date of filing does not establish priority. After public disclosure, designs and devices that have not been patented are open for anyone's use. This is quite different from the status of research after public disclosure. No strict limits apply to the time lapse between reporting on one's research at scientific meetings and publishing a detailed report of the work, although as we saw in the scenario "A Premature Claim" in Chapter 6, announcing a result that one does not deliver can cause problems for others and diminish one's standing in a field.

INDIVIDUAL CREDIT AND THE OWNERSHIP OF INNOVATION

The framework of laws and conventions covering the fair use of intellectual property may itself be questioned. The view that people should freely share their good ideas found strong advocates among, for instance, the Shakers, a celibate religious sect that flourished in the United States in the nineteenth century. The Shakers believed that the ways they found to reduce drudgery and make work an expression of love should be freely offered to the world. Thus have the clothespin and the flat-broom become part of Americana. The names of those who invented the remarkable number and range of Shaker devices (farm implements, household objects, furniture, and clothing designs) remain largely unknown. For decades, Shakers refused to take out patents. Only when outsiders began to patent Shaker inventions did this community apply for them themselves. Even when they did not patent their innovations, the Shakers felt they profited by using their devices and by selling them.

Whether or not you find this Shaker outlook compelling, it is an alternative to the assumption that competitive advantage in the market is the prime value. How broadly should one share ideas? How readily should one copy the ideas of others? How much does it matter what the ideas are?

What about a surgeon who develops a technique that can save lives but keeps the technique a "trade secret" to enhance his relative standing?

Is withholding the technique morally wrong? Would it be wrong for another surgeon to try to learn that technique, by, say, electronic eavesdropping or asking an operating room assistant? Are the ethical limits on publicizing another's surgical innovation the same as for other sorts of innovation?

These questions raise issues concerning the public's welfare. Does the present system of intellectual property ownership and social control ensure that members of the professions serve the public interest? We saw in the introduction that the rationale for the U.S. constitutional provision granting authors and inventors a time limit for exclusive legal right to their writings, discoveries, and inventions was to stimulate creativity of such works *for the benefit of society.* In the case of patents, the temporary exclusive right gives inventors an incentive to make their innovations patent, publicly known, rather than keeping them secret.

BENCHMARKING AND REVERSE ENGINEERING

Beyond the issue of respecting patents, what issues do copying and learning from the innovations of others raise? The problem is not simply to prevent unfair copying. It is also to prevent the "not invented here" syndrome, the refusal to learn from and recognize the value of others' innovations, which frequently prevents engineers from attaining achievable levels of quality and safety in new products.

A commonly accepted first step in the design process is benchmarking. The ordinary English sense of "benchmark" is a standard by which a thing can be measured. In engineering, "benchmarking" refers to obtaining a competitor's devices or information before one designs and manufactures a new product. If the product is not prohibitively expensive – not, for example, a nuclear power plant – samples of the competitor's product are commonly purchased, examined, and analyzed. Benchmarking may or may not involve copying anything from the competitor. A company might wish to examine its competitors' products or pricing structures to learn about the competitors' cost of manufacture so as to judge whether, with some new manufacturing process, the company can enter the market and produce a competitive product at a much lower price.

One means of obtaining information about a competitor's product is reverse engineering. Reverse engineering is the examination of another company's product to understand the technology and process used in its design, manufacture, or operation. It usually involves taking the product apart and testing ways to destroy it. Reverse engineering is commonly used to learn what a competitor has done to copy or improve on it. For example, engineers might photograph and enlarge pictures of silicon chips to learn about the architectural features of the chips such as whether to use one function twice or two different functions once. (They would not copy every detail of the chip. This would be self-defeating since then their company would bring out the same product later than competitors.)

How should we assess such copying? In reverse engineering, no credit is given to the individuals who originate the innovations or the company that employs them. Is this duplication a form of plagiarism, the copying of another's words or ideas? Copying a design might also violate a patent or a copyright, and that would leave one open to a lawsuit. The question of plagiarism is not, however, a question of legality. We saw in Chapter 9 that plagiarism (as contrasted with copyright infringement) is not illegal. It is, nevertheless, wrong and entails penalties, such as expulsion from school and debarment from grant competitions.

Nonetheless, adhering to the scholarly standard of citation and acknowledgment in every aspect of life, and attempting to acknowledge others for everything we learn from them, would absurdly burden life. The mechanisms for acknowledging others' engineering innovations are limited to obtaining licenses for copying materials and designs covered by copyright and patent and to respecting confidentiality agreements. Is copying without acknowledgment of anything that is not so protected therefore ethically permissible? Does the answer extend to trade secrets?

The ethical constraints that are generally recognized in benchmarking and reverse engineering (other than respecting legal property rights) are constraints on the means one can use to obtain information rather than the nature of the information or the use one makes of it.

In Chapter 5 we examined some guidance that the Texas Instruments Ethics Office gives its employees. What it has to say about benchmarking and reverse engineering is instructive for what it shows about where an ethically concerned company draws the line in these matters. The office lists the following acceptable benchmarking practices:

- Asking customers about equipment and prices of TI competitors.
- Asking employees of well-run businesses that don't compete with TI about their practices.
- Searching for information through public resources.
- Reading books and publications describing other companies.

- Encouraging other TIers who come in contact with customers to be observant of practices that might be useful to TI.

Practices Texas Instruments excludes as unethical include:

- Misrepresenting oneself as working for another employer.
- Collusion in fixing prices or allocating markets or customers.
- Disparaging a competitor's business to customers or to others.
- Attempting to gain confidential information about other businesses.[2]

Misrepresentation is deception and thus as morally objectionable as lying. Collusion is a secret agreement for a deceitful or a fraudulent purpose. Collusion to fix prices is illegal. Finding out about the competition's pricing structure might lead to price fixing, which is more of a temptation for companies than for engineers. The danger of appearing to fix prices is relevant to the topic of claiming credit for engineering work. This threat is one reason TI tells its employees never to attempt gaining competitive information directly from a competitor. Indeed, companies may purchase benchmarking from others so that the benchmarking will not involve direct communication with competitors.

Disparaging a competitor to customers may be a poor policy rather than unethical, assuming you are truthful in your assessment of the competitor's failings. Like negative campaigning by a politician, disparagement is likely to lead people to believe that you focus on the competitor's failing because you have little positive to offer, or that your whole industry is corrupt or incompetent.

The first three objectionable practices are relatively easily recognized, but the line between the fourth, attempting to gain confidential information and the acceptable practice of obtaining information through public resources, is somewhat harder to discern. What about going into the showroom of a competitor, as though one were a customer, and asking questions of the salesperson? TI holds that it is not necessary to tell the salesperson the name of one's employer or the reason for one's interest. (This judgment accords with the practice of many other reputable companies.)

In this situation we see the standard of disclosure required for honesty from an engineer engaged in benchmarking and that it is rather different from the standard of disclosure required of, say, a journalist getting a story. It is wrong for either the engineer giving a professional opinion or the journalist writing a story to tell misleading half-truths, of course, but now we are considering how much each must disclose to third parties from whom they seek to get the information necessary to do their jobs. A journalist who does not disclose that he is a journalist will be judged for behaving deceptively, tricking others into speaking more candidly than they would if they knew their remarks were to be reported. An engineer engaged in benchmarking who does not disclose to a competitor's salesperson what she is doing is seen as acting deceptively only if she lies to the salesperson.

As we saw in Chapter 2, what is entrusted to engineers is different from what is entrusted to members of other professions. Understanding what is central to fulfilling the public's trust in each profession is a complex matter. The provisions in ethical codes and guidelines are justified insofar as they express what is necessary or important to fulfilling that trust.

Consider whether it is ethically permissible for a company to send a product obtained from one supplier to a second supplier so that the second supplier can reverse engineer it. If the first supplier had made a confidentiality agreement, clearly the action would be illegal. If there were no confidentiality agreement, would the action be ethically permissible? The accepted practice is for the second supplier to purchase the product from the other supplier (or distributor) and then to use reverse engineering to find out about the competitor's product. It is suspicious and perhaps unfair to the first supplier for the customer to play an active role in the second supplier's attempt to reverse engineer the product.[3] If the customer had some special advantage in obtaining the product or information about it, the action would certainly be unfair.[4]

CONCLUSION

In this chapter we have examined fair credit primarily in terms of prevailing laws and customs in technologically developed democracies with market economies. We have also reflected briefly on the legitimacy of those laws and customs by comparing them with the practices of a community that once produced many technological innovations but measured success in terms other than market competition.

The laws and standards now covering technological innovation show that learning from others, even in the absence of any explicit means of according them credit, is the norm in technological innovation so long as one meets ethical constraints in the process of acquiring information and does not violate copyrights, patents, or trademarks.

NOTES

[www]Based on NSPE Board of Ethical Review Case 83-3.

[www2]Based on NSPE. Board of Ethical Review Case 77-5.

[www3]Adapted from a scenario by Stephen A . Ward, based on current legal cases. This scenario is the basis of a project on reverse engineering by John Wallberg that may be found in the WWW Ethics Center for Engineering and Science.

1. For an extensive discussion of intellectual property rights, see Weil, Vivian and John W. Snapper. 1989. *Owning Scientific & Technical Information: Value and Ethical Issues.* New Brunswick NJ: Rutgers University Press.

2. Article Number 72 from the TI Ethics Office, available in the Ethics in a Corporate Setting section of the WWW Ethics Center for Engineering and Science.

3. This is the TI Ethics Office's assessment of the situation. See TI Ethics Office Article 142, Reverse engineering and patent infringement, in the WWW Ethics Center for Engineering and Science.

4. *Ibid.*

EPILOG

Making a Life in Engineering and Science

In this book we have examined many aspects of professional responsibility for engineers and scientists. Many aspects of moral life lie outside considerations of professional responsibility. Family responsibility and general civic responsibility are two other major areas of moral responsibility that clearly lie outside the scope of considerations of this book. What about the choice of work in engineering and science? Such decisions are made within the context of many other decisions, including family obligations. For example, family obligations may restrict the geographical region in which you seek work. Practical considerations such as the need to pay back education loans also influence job choices. The choice of work is more intimately connected to professional ethics, because it significantly influences the opportunities for expressing one's values in one's work. Work that fulfills one's aspirations as well as ambitions and need for income is a major element in a meaningful life. How does one find opportunities to do such work? This is a problem, indeed a design problem, that a person addresses many times in a life, if at all. (I say "if at all," because many people in the world today and throughout human history have had little opportunity to pursue many aspirations in their work life beyond providing subsistence to themselves and their families.)

The current range of possibilities for a young adult with talent in engineering and science may itself be daunting, and it would only add to that burden to attempt to catalog the value dimensions of work choice. In any case I would be reluctant to do so, since I have too often seen humanists and social scientists much more ready to instruct engineers and scientists about the goals they should pursue than to consider the social implications of their own work in humanities and social science. I would not want to risk adding to that abuse.

Instead, I will tell you about two of the many engineers it has been my pleasure to know who have acted on their aspirations. The stories

306

I shall recount are about two engineers at very different stages of their career and making contributions of very different sorts based on quite different concerns and priorities. Your own values may or may not be like either of the two engineers in the stories I tell here.

MIGUEL BARRIENTOS, BUILDING A WATER PUMP FOR ANDEAN ALPACA BREEDERS

First is a story of an undergraduate project carried out by Miguel Barrientos, MIT '93, to design and manufacture a human-powered treadle pump to meet the needs of Andean alpaca breeders.[1] Miguel found the project a fulfilling experience. It enabled him to gain valuable experience in the field of appropriate technology – that is, the development of technology suited to the needs of small producers, rural and urban, especially in the developing world. It was personally rewarding because it enabled him to work for the betterment of his native country, Bolivia.

As Miguel knew, people who live in rural areas of Bolivia continually face the problem of supplying water for their houses, crops, and animals. This problem became particularly acute immediately preceding his project because of a drought in the Andean regions of the country that had begun in 1989. The drought had become a major obstacle to most of the development projects that operate in the Bolivian Andes.

From June 8 to September 4, 1992, Miguel worked as a technical consultant for the Alpaca Wool Production and Processing Project (PPPLA). The PPPLA endeavors to improve breeding and veterinarian practices among alpaca breeders in the Andean zones of Bolivia; it receives funding from the United Nations Development Program (UNDP), the United Nations Capital Development Fund (UNCDF), and Appropriate Technology International (ATI)[www]. Miguel was asked to identify an appropriate type of water pump and assist with its production. The pump had to be suitable for use with a water table two to five meters deep, inexpensive to manufacture, and simple to maintain and repair. ATI had previously tested a human-powered treadle pump for use in Africa. The area for the intended use of the pump in Bolivia is cold and dry because of the high altitude of the zone (4,000 to 5,000 meters above sea level).

Miguel began his project by visiting with some of the alpaca breeders for whom the pump was to be built to better understand their needs and conditions of the pump's use. The design considerations Miguel identified for the treadle water pump were that it be

- versatile – its design should suit it for use to feed livestock, irrigate pastures and crops, and even provide water for households.
- inexpensive – sell for about 80 U.S. dollars. (Existing pumps sold for at least 140 U.S. dollars.)
- reparable by the alpaca breeders themselves.

The project needed skilled artisans in Bolivia to build the pump for the Andes. Miguel located a small machine shop called "Khana Wayra" (which in the Aymara language means "Light of the Wind") whose owners were experienced in producing wind turbines, manual and wind-powered water pumps, solar heaters, and drilling equipment. With the members of Khana Wayra, Miguel built and tested a prototype treadle pump. By July the pump was ready for a demonstration trip to the project area.

Miguel modified the original design of the treadle pump in several ways: Cylinders made of PVC pipe replaced the cylinders made of sheet steel to make it easier to maintain. He enlarged the valve box and valve plate to accommodate PVC cylinders, which were slightly larger than the metal cylinders. He redesigned the inlet pipe and the treadle support so that they could be detached from the main body to make the pump easier to transport.

Khana Wayra's machinists were experienced in building water pumps, so they completed the prototype pump in four days.

Miguel and Pablo Garay, a member of Khana Wayra, tested the prototype pump briefly in early July, satisfactorily pumping water at floor level. Several days later when they took the pump to a well, they found that the outlet valves were not making a tight seal, allowing air to be sucked into the pump. The leakage was due to the imperfect roundness of the cylinders and to the poor seal made by the rubber disk they were using. Khana Wayra did not have a lathe where the cylinders could be turned down, so the only way to improve suction was to replace the rubber disk. They molded leather cups to replace the rubber disks, treated them with vegetable oil, and tested the pump again. They successfully pumped water from a depth of 1.5 meters, but suction was too weak to pump from a depth of 3.5 meters. They then replaced the inlet pipe, which had a diameter of 1.5 inches, with a 1-inch-diameter pipe, and tested the pump again. The pump functioned well, pumping water at a rate of approximately 1.3 liters per second.

With the machinists of Khana Wayra Miguel then determined the maximum depth at which the pump could function with a 1.5-inch-diameter pipe. They pumped water successfully first from 2.5 meters, then from 3 meters, and finally from 4.8 meters. The pump operated flawlessly even in the 4.8-meter well, extracting water at a rate of approximately 0.5 liters per second. It did not fully fill the cylinders in a 6-meter well, but since the theoretical depth limit for a treadle pump at 4,000 meters above sea level (atmospheric pressure = 470 mm Hg) is 6.4 meters, they were satisfied with the performance of the prototype.

Next came field testing. The field testing had several purposes:

1. To demonstrate the characteristics of the treadle pump to the alpaca breeders.
2. To test the pump in the area where it was to be used.
3. To identify flaws of the pump design, both through tests and feedback from the potential users.

At the first site, Miguel first tried the pump in one location where water can be found at a depth of 1.5 meters, but found the water too muddy to pump efficiently. In that location, observers were disappointed in the amount of physical labor required to extract the water. Miguel then tried the pump at a new location, at which he found clear water at a depth of 2 meters. One of the spectators volunteered to try it out. The pump operated flawlessly, pumping water at a rate that impressed the people watching the demonstration.

The field testing and comments from local observers at Cosapa and later at Wariscata led Miguel to recommend the following changes before the pump went into mass production:

1. Increase support to the treadles' axle to make the pump structure more rigid.
2. Clearly mark the inlet pipe at the point of attachment of the treadle support pipe, to facilitate assembly.
3. Construct the treadles of 4 by 2 inch hardwood to make the operation of the pump safer.
4. Attach the treadle support pipe to the inlet pipe so that the pump operator does not have to get too close to the well.
5. Widen the baseboard to increase the stability of the pump.
6. Increase the thickness of the pulley's rope because the rope wears rapidly.

7. Use stainless steel springs in the outlet valves to prevent corrosion, and modify the valves to eliminate leaks.
8. Glue the rubber seal of the valve box to the baseboard to avoid deformation caused by suction.
9. Use special tooling to accelerate the production rate of treadle pumps. (This tooling was subsequently received from ATI.)
10. Provide blueprints and instructions to those observers who expressed interest in producing some of the parts themselves.

Along with these recommendations, Miguel gave Khana Wayra new blueprints of the pump incorporating all the modifications to the original pump design and instructed Khana Wayra's members on the use of the tooling they received from ATI.

JIM MELCHER, WITNESSING AGAINST WASTE AND VIOLENCE

At his untimely death of cancer at age 54, James R. Melcher was the Julius A. Stratton Professor of Electrical Engineering and Physics and director of MIT's Laboratory for Electromagnetic and Electronic Systems (LEES), one of the large interdisciplinary laboratories at MIT. He was widely known in electrical engineering for his practical applications of continuum electromechanics, a broadly interdisciplinary field that draws on electromagnetics, fluid and solid mechanics, heat transfer, and physical chemistry. His strong interest in the ethical questions raised by engineering work was long evident to his colleagues in the MIT Electrical Engineering and Computer Science Department and around the world.

The son of a Methodist minister, Jim was a kind and modest person, with the courage to look unflinchingly at very difficult problems. He did not shrink from considering any implications of engineering activities. I recall one of the Institute Professors at MIT who had known Jim since graduate school describing Jim as a "saint."

A key experience for Jim in becoming a vocal critic of militarism and energy dependence had been his sabbatical year at the Cavendish Laboratory at Cambridge University in 1971 where he was working on his text, *Continuum Electromechanics*. He described the effect of seeing the United States from abroad as convincing him to have a stronger influence on the course the United States was taking.[2]

The first Arab oil boycott took place shortly after Jim's return to the United States. The experience of the effects of this boycott together with his growing awareness of the United States energy vulnerability led him to undertake a striking witness, to give up his second car and make his daily commute, 18-miles round trip, by bicycle. Jim had always gotten plenty of exercise, in part to control his diabetes, but with university athletic facilities at his disposal he had many exercise options that were more pleasant than biking 18 miles a day through Boston winters. Bicycle racing was one of the side benefits that attracted graduate students to work with Jim in the Laboratory for Continuum Electromechanics, which he had founded within LEES.

Jim's research now became more applied: limiting air pollution from diesel exhaust and coal cumbustors. His students, too, took their theoretical work on topics such as the mass transfer of electric fields in fluidized beds and immediately applied them to making environmentally friendly ways to recycle asphalt concrete.

Jim recounts his realization that "For the sake of oil, our government cast its lot with an obscenely rich dictatorship [in Iran] which was out of step with popular movements." He particularly recalled an interview with the then shah of Iran on a U.S. news program. The shah, whose dictatorship is widely acknowledged to have been supported by the CIA, unblinkingly affirmed he was God. Jim observed that had the U.S. government given direct subsidies to U.S. corporations to purchase oil at even $2/gallon, it would have been cheaper in financial terms than trying to keep oil "cheap" by military force.

> Without a hidden military subsidy of foreign oil, domestic oil would be competitive. Embedded in shale, for example, there is more oil in the United States than in the Middle East. Shale oil would be competitive if the price of oil were little higher than it is now – and if that price held steady as crises came and went.

He tried unsuccessfully to advance the exploration of these domestic alternatives.

Jim was additionally appalled at what he saw as the over-readiness of the United States to resort to military means or CIA tactics in support of unpopular dictatorships. Jim found the quick resort to violence in support of materialism quite inconsistent with Christianity as he understood it.

When President Reagan announced the Strategic Defense Initiative, or Star Wars, in 1983, Jim, like many of his peers, saw this as much a political initiative as a technical one; he was concerned about what he saw as the Pentagon's efforts to make university pursuit of SDI funds look like an endorsement of the project, which many believed was fundamentally flawed and promising only to increase a growing national deficit. The talents of LEES were ideally suited to SDI research, and Jim made it clear to the faculty in his laboratory that he would not block any proposals for SDI funding from individual faculty in his lab, but he sought to find other sources of support and was quite successful in doing so. (At this time engineering salaries had yet to undergo "hardening," and many universities, including Jim's, expected their engineering faculty to raise a great deal of grant money and even to cover half of their own salary by this means.)

Personally, Jim wanted to do more, so he became part of a campaign to convince senior faculty in engineering and science departments around the country to pledge not to take SDI funding. On May 13, 1986, he joined with three Nobel laureates at a Washington press conference to make public over 3,700 names of those who had pledged not to take SDI funds.

In his last months, as Jim squarely faced his losing bout with cancer, he turned to writing his experience of making larger sense of things as he made a life in engineering. I leave you with some of his words from that article:

> To really integrate the way you earn your living with your social and even spiritual aspirations, for people in any line of work, is the true test of an education. Your values must become part of your professional thinking which is best learned "hands on."[2]

NOTES

1. This account is primarily based on the final report on the project "Designing Tools for Developing Countries" written by Miguel Barrientos, MIT '93, and my discussions with him about the project.
2. "America's Perestroika, Living a New National Agenda" by Professor James R. Melcher, PhD '62, *Technology Review*, April 1991, MIT, pp. 4–11. This article, my own recollections, and the obituary for Jim published in the January 9, 1991 issue of *Tech Talk* are the principal sources for the present account.

BIBLIOGRAPHY AND REFERENCES

Addelson, Kathryn Pyne. 1991. *Impure Thoughts: Essays on Philosophy, Feminism and Ethics*. Philadelphia: Temple University Press.

Addelson, Kathryn. 1994. *Moral Passages*. New York: Routledge, pp. 13–18.

Alberts, Bruce and Kenneth Shine. 1994. "Scientists and the Integrity of Research." *Science*, 266 (December 9), 1660.

Allen, Anita L. 1987. *Uneasy Access: Privacy for Women in a Free Society*. Totowa, NJ: Rowman and Allanheld.

——— 1995. "Privacy in Health Care." In the second edition of the *Encyclopedia of Bioethics*. New York: Macmillan, pp. 2064–2073.

American Chemical Society. 1994. "Ethical Guidelines to Publication of Chemical Research." *Acc. Chemical Research*, 27(6):179–181.

Ashford, Nicholas A. 1986. "Medical Screening in the Workplace: Legal and Ethical Considerations." *Seminars in Occupational Medicine*, 1(1):67–79.

Ashford, Nicholas A., Carla Bregman, Dale B. Hattis, Abyd Karmali, Christine Schabacker, Linda-Jo Schierow, and Caroline Whitbeck. *Monitoring the Community for Exposure and Disease: Scientific, Legal, and Ethical Considerations*, a report supported by the Agency for Toxic Substance and Disease Registry (ATSDR) and the National Institute for Occupational Safety and Health (NIOSH), 1991.

Austin, J. L. 1961. "A Plea for Excuses." In *Philosophical Papers*, pp. 123–152, edited by J. O. Urmson and G. J. Warnock. London: Oxford University Press.

Babbage, Charles. 1830. *Reflections on the Decline of Science in England in Science and Reform: Selected Works of Charles Babbage*. Cambridge; New York: Cambridge University Press, 1989.

Baier, Annette. 1986a. "Extending the Limits of Moral Theory." *The Journal of Philosophy*, 77:538–545.

——— 1986b. "Trust and Antitrust." *Ethics*, 96:232–260. Reprinted in *Moral Prejudices*, pp. 95–129. Cambridge, MA: Harvard University Press.

——— 1990. "A Naturalist View of Persons." Presidential Address delivered before the Eighty-Seventh Annual Eastern Division Meeting of the American Philosophical Association in Boston, Massachusetts, December 29, 1990. *APA Proceedings*, 65(3):5–17.

——— 1993. "Claims, Rights, Responsibilities." In *Prospects for A Common Morality*, edited by J. P. Reeder and G. Outka, Princeton University Press, and in a collection of

Baier's essays titled *Passions of the Mind*, forthcoming from Harvard University Press.

Barber, Bernard. 1983. *The Logic and Limits of Trust.* New Brunswick, NJ: Rutgers University Press.

Baron, Marcia. 1984. *The Moral Status of Loyalty.* Dubuque, Iowa: Kendall/Hunt Publishing Co.

Beardsley, Tim. 1996. "Profile: Thereza Imanishi-Kari." *Scientific American,* 275(5) (November):50–52.

Beauchamp, Tom and James Childress, editors. 1979. *Principles of Biomedical Ethics.* London: Oxford University Press.

Beecher, Henry K. 1966. "Ethics and Clinical Research." *New England Journal of Medicine,* 274(24) (June 16, 1966):1354–1360.

Benner, Patricia. 1984. *From Novice to Expert: Excellence and Power in Clinical Nursing.* Redding, MA: Addison-Wesley.

Benner, Patricia and Judith Wrubel. 1989. *The Primacy of Caring: Stress and Coping in Health and Illness.* Reading, MA: Addison-Wesley.

Binder, S. and S. Bonzo. 1989. "Letter to the Editor." *American Journal of Public Health,* 79(12):1681.

Bird, Stephanie J. and David E. Housman. 1995. "Trust and the Collection, Selection, Analysis and Interpretation of Data: A Scientist's View." *Science and Engineering Ethics,* 1(4) (October):371.

Bird, Stephanie J. and Jerome Rothenberg. 1988. "To Screen or Not to Screen: Drugs, DNA, AIDS." Unpublished manuscript.

Block, Peter. 1993. *Stewardship – Choosing Service Over Self-Interest.* San Francisco: Berrett-Koehler Publishers.

Board of Ethical Review of the National Society of Professional Engineers. 1976. *Opinions of the Board of Ethical Review,* Volume IV. Alexandria, Virginia: National Society of Professional Engineers.

1981. *Opinions of the Board of Ethical Review,* Volume V. Alexandria, Virginia: National Society of Professional Engineers.

1989. *Opinions of the Board of Ethical Review,* Volume VI. Alexandria, Virginia: National Society of Professional Engineers.

1994. *Opinions of the Board of Ethical Review,* Volume VII. Alexandria, Virginia: National Society of Professional Engineers.

Bosk, Charles L. 1979. *Forgive and Remember: Managing Medical Failure.* Chicago: University of Chicago Press.

Brandt, Allan M. 1985. *No Magic Bullet: A Social History of Venereal Disease in the United States Since 1880.* New York: Oxford University Press.

Briggs, Shirley A. 1987. *Rachel Carson: Her Vision and Her Legacy.* In *Silent Spring Revisited.* Marco, Gino J., Hollingworth, Robert M., Durham, William (eds.), pp. 3–11. Washington DC: American Chemical Society.

Broad, William J. and Nicholas Wade. 1982. *Betrayers of the Truth.* New York: Simon and Schuster.

Broome, Taft H., Jr. 1986. "The Slippery Ethics of Engineering." *The Washington Post.* December, 28.

Bucciarelli, Louis L. 1985. "Is Idiot Proof Safe Enough?" *Applied Philosophy,* 2(4): 49–57; reprinted in *Ethics and Risk Management in Engineering,* edited by

Albert Flores. Landam, New York & London: University Press of America, pp. 201–209.

Bullard, R. D. 1990. *Dumping in Dixie: Race, Class and Environmental Quality.* Boulder, CO: Westview Press.

Buzzelli, Donald E. 1993. "The Definition of Misconduct in Science: A View from NSF." *Science,* 259:584–648.

Callahan, Daniel, Arthur Caplan, and Bruce Jennings (eds.). 1985. *Applying the Humanities.* New York: Plenum.

Caplan, Arthur L. 1992. *When Medicine Went Mad: Bioethics and the Holocaust.* Totowa, NJ: Humana Press.

Cohen, Jon. 1994. "U.S.–French Patent Dispute Heads for a Showdown." *Science* 265 (July 1, 1994):23–25.

Committee on Academic Responsibility Appointed by the President and Provost of MIT. 1992. *Fostering Academic Integrity.* Boston: Massachusetts Institute of Technology.

Committee on Editorial Policy of Council of Biology Editors [CBE]. 1990. *Ethics and Policy in Scientific Publication.* Bethesda, Maryland: Council of Biology Editors, Inc.

Committee on Engineering Design Theory and Methodology, the Manufacturing Studies Board of the National Research Council. 1991. *Improving Engineering Design: Designing for Competitive Advantage.* Washington DC: National Academy Press.

Committee on Science, Engineering and Public Policy of the National Academy of Sciences, National Academy of Engineering, and Institute of Medicine. 1995. *On Being a Scientist,* second edition. Washington, DC: National Academy Press.

Culliton, Barbara J. 1988. "Harvard Tackles the Rush to Publication." *Science,* 241 (July 29, 1988):525.

Curd, Martin and Larry May. 1984. *Responsibility for Harmful Actions.* Dubuque, Iowa: Kendall/Hunt Publishing Co.

Dandekar, Natalie. 1991. "Can Whistleblowing be Fully Legitimated?" *Business and Professional Ethics Journal,* 10(1):89–108.

Davis, Michael. 1988. "Avoiding the Tragedy of Whistleblowing." *Business & Professional Ethics Journal,* 8(4):3–19.

1991. "Avoiding the Tragedy of Whistleblowing." *Business and Professional Ethics Journal,* 8(4):3–19.

1997. "Better Communications Between Engineers and Managers: Some Ways to Prevent Many Ethically Hard Choices." 1997. *Science and Engineering Ethics,* 3(2):in press.

Djerassi, Carl. 1989. *Cantor's Dilemma.* New York: Penguin Books, USA.

1994. *The Bourbaki Gambit.* Athens: University of Georgia Press.

Dowie, Mark. 1977. "Pinto Madness." *Mother Jones* (September/October):19–32.

Eads, George, and Peter Reuter. 1983. Designing Safer Products. The Rand Institute for Civil Justice.

Eddy, Paul, Elaine Potter, and Bruce Page. 1979. "Is the DC-10 a Lemon?" *New Republic,* (June 9):7–9.

Elliston, Frederick Keenan, Lockhart, van Schaick. 1985. *Whistleblowing: Managing Dissent in the Workplace.* New York: Praeger Scientific.

Elstein, Arthur S., Lee S. Shulman, and Sarah A. Sprafka. 1978. *Medical Problem Solving: An Analysis of Clinical Reasoning.* Cambridge, MA: Harvard University Press.

Flumerfelt, R. W., C. E. Harris, Michael J. Rabins, and C. H. Samson. 199?. *Engineering Ethics* (Texas A&M), final report to the NSF on Grant Number DIR-9012252.

French, Peter. 1982. "What is Hamlet to McDonnell-Douglas or McDonnell-Douglas to Hamlet: DC-10." *Business and Professional Ethics Journal,* 1(2):1–14.

Friedman, Milton. 1970. "The Social Responsibility of Business Is To Increase Its Profits." *The New York Times Magazine* (September 13); reprinted in *Ethical Issues in Engineering,* 1991, edited by Deborah Johnson, Englewood Cliffs, NJ: Prentice-Hall, pp. 78–83.

Frye, Marilyn. 1983. *The Politics of Reality: Essays in Feminist Theory.* Trumansburg, NY: The Crossing Press.

Gibbs, Lois M. 1982. *Love Canal: My Story.* Albany, NY: State University of New York Press.

1985. *Centers for Disease Control: Cover-up, Deceit and Confusion.* Arlington, VA: Citizens' Clearinghouse for Hazardous Wastes.

Gilbane Gold. 1989. 24-minute videotaped dramatization produced by the National Institute for Engineering Ethics, National Society of Professional Engineers.

Gilligan, Carol. 1982. *In a Different Voice: Psychological Theory and Women's Development.* Cambridge, MA: Harvard University Press.

Goleman, Daniel. 1985. *Vital Lies, Simple Truths.* New York: Simon & Schuster, Inc.

Goodstein, David L. 1994. "Whatever Happened to Cold Fusion?" *Engineering & Science,* Fall 15–25; reprinted from *the American Scholar,* 63:4 Autumn 1994.

1995. "Ethics and Peer Review." *Biotechnology,* 13 (June 1995):618.

Graham, Loren. 1993. *The Ghost of the Executed Engineer: Technology and the Fall of the Soviet Union.* Cambridge, MA: Harvard University Press.

Gunn, Alastair and P. Aarne Vesilind. 1983. "Ethics and Engineering Education." *Journal of Professional Issues in Engineering,* ASCE. vol. 109, No. 2.

1986. *Environmental Ethics for Engineers.* Chelsea, MI: Lewis Publishers.

1990. "Why Can't You Ethicists Tell Me the Right Answers?" *Journal of Professional Issues in Engineering,* ASCE, vol. 116, No. 1.

Hampshire, Stuart. 1949. "Fallacies in Moral Philosophy." *Mind,* 58:466–482; reprinted in *Revisions: Changing Perspectives in Moral Philosophy,* edited by Stanley Hauerwas and Alasdair MacIntrye, 1983.

1989. *Innocence and Experience.* Cambridge, MA: Harvard University Press.

Harris, Charles E., Jr. 1991. "Manufacturers and the Environment: Three Alternative Views." In *Environmentally Conscious Managing: Recent Advances,* edited by Mo Jamshidi, Mo Shahinpoor, and J. H. Mullins. Albuquerque: ECM Press.

Harris, Charles E., Jr., Michael S. Pritchard, and Michael J. Rabins. 1995. *Engineering Ethics.* Wadsworth.

Hauerwas, Stanley. 1977. *Truthfulness and Tragedy.* Notre Dame, IN: University of Notre Dame Press.

1983. "Constancy and Forgiveness, The Novel as School for Virtue. "*Notre Dame Literary Journal,* 15(3).

Hauerwas, Stanley and David Burrell. 1977. "From System to Story: An Alternative Pattern for Rationality in Ethics." In *Truthfulness and Tragedy*, pp. 15–39; reprinted in *Why Narrative?*, 158–190.

Hauerwas, Stanley, and L. Gregory Jones, (eds.). 1989. *Why Narrative?* Grand Rapids, MI: William Eerdmans.

Hauerwas, Stanley and Alasdair, MacIntrye, (eds.). 1983. *Revisions: Changing Perspectives in Moral Philosophy.* Notre Dame, IN: University of Notre Dame Press.

Hick, John. 1993. *Disputed Questions in Theology and the Philosophy of Religion.* New Haven, CT: Yale University Press, p. 93.

Hoerr, John, with William G. Glaberson, Daniel B. Moskowitz, Vicky Cahan, Michael A. Pollock, and Jonathan Tasini. 1985. "Beyond Unions." *Business Week*, July 8.

HHS Commission on Research Integrity. 1995. "Professional Misconduct Involving Research." *Professional Ethics Report*, vol. VIII, no. 3 (Summer '95).

Holton, Gerald. 1978. "Subelectrons, Presuppositions, and the Millikan-Ehrenhaft Dispute." In *Historical Studies in the Physical Sciences*, 11:166–224; reprinted in the collection of Holton's essays, *Scientific Imagination*, Cambridge, MA: Cambridge University Press, 1978, pp. 25–83.

1994. "On Doing One's Damnedest: The Evolution of Trust in Scientific Findings." Chapter 7 in Holton's *Einstein, History, and Other Passions*. New York: American Institute of Physics.

Hurst, J. Willis and H. Kenneth Walker, (eds.). 1972. *The Problem-Oriented System.* Baltimore: The Williams and Wilkins Company.

Jackall, Robert. 1988. *Moral Mazes.* New York: Oxford University Press.

Jackson, C. Ian. 1986. *Honor in Science.* New Haven, CT: Sigma Xi, the Scientific Research Society.

Janis, Irving. 1982. *Groupthink*, 2nd edition. Boston: Houghton Mifflin.

Jasanoff, Sheila S. 1985. *Controlling Chemicals: The Politics of Regulation in Europe and the U.S.* Ithaca: Cornell University Press.

Jonsen, Albert R. 1977. "Do No Harm: Axiom of Medical Ethics." In *Philosophical Medical Ethics: Its Nature and Significance*, edited by S. F. Spicker and H. T. Engelhardt, Jr., Dordrecht, Holland: D. Reidel Publishing Co.

1991. "Of Balloons and Bicycles, or the Relationship between Ethical Theory and Practical Judgment." *Hastings Center Report*, 21(5):14–16.

Jonsen, Albert R. and Stephen Toulmin. 1988. *The Abuse of Casuistry: A History of Moral Reasoning.* Berkeley: University of California Press.

Kaiser, Jocelyn. 1995. "Commission Proposes New Definition of Misconduct," *Science*, 269 (29 September, 1995):1811.

Kaufman, Ron. 1992. "After 5 Years, Heated Controversy Persists in Science Copyright Case." *The Scientist*, September 14, 1992, 1, 4, 5,10.

Kiang, Nelson. 1995. "How are Scientific Corrections Made?" *Science and Engineering Ethics*, 1, 4(October):347.

Korenman, Stanley G. and Allan C. Shipp with Association of American Medical Colleges ad hoc Committee on Misconduct and Conflict of Interest. 1994. *Teaching the Responsible Conduct of Research Through a Case Study Approach.* New York: Association of American Medical Colleges.

Knox, Richard A. 1993. "Care at the End Not as Costly as Assumed." *The Boston Globe*, 243(105):A1, April 14.

Kuhn, Thomas S. 1977. *The Essential Tension.* Chicago: University of Chicago Press.

Lachman, Judith A. 1990. *Issues in Management, Law and Ethics,* Chapter 22, unpublished manuscript.

Ladd, John. 1970. "Morality and the Ideal of Rationality in Formal Organizations." *The Monist,* 54(4):488–516.

————. 1975. "The Ethics of Participation." In *Participation in Politics: NOMOS XVI,* pp. 98–125. New York: Atherton-Leiber.

————. 1976. "Are Ethics and Science Compatible?" In *Science, Ethics, and Medicine,* pp. 49–78. Hastings-on-Hudson: The Hastings Center.

————. 1979. "Legalism and Medical Ethics." In *Contemporary Issues in Biomedical Ethics,* pp. 1–35. Clifton, NJ: Humana Press.

————. 1982a. "The Distinction Between Rights and Responsibilities: A Defense." *Linacre Quarterly,* 49:121–142.

————. 1982b. "Philosophical Remarks on Professional Responsibility in Organizations." *Applied Philosophy,* 1(Fall):58–70.

————. "The Quest for a Code of Professional Ethics: An Intellectual and Moral Confusion." In *AAAS Professional Ethics Project: Professional Ethics Activities in the Science and Engineering Societies,* Rosemary Chalk, Mark S. Frankel, and Sallie B. Chafer. Washington, DC: AAAS Press. Reprinted in *Ethical Issues in Engineering,* edited by Deborah Johnson. Englewood Cliffs, NJ: Prentice-Hall, 1990, pp. 130–136.

Layton, Edwin T. 1985. "Theory and Application in Science and the Humanities." In *Applying the Humanities,* edited by Daniel Callahan, Arthur L. Caplan, and Bruce Jennings. New York: Plenum Press, pp. 57–70.

Leveson, Nancy G. and Clark S. Turner. 1993. "An Investigation of the Therac-25 Accidents." *Computer* (published by IEEE) (July 1993), pp. 18–41.

Levine, Robert J. 1986. *Ethics and the Regulation of Clinical Research.* 2d ed. Baltimore: Urban & Schwarzenberg.

————. 1995. "Informed Consent: Consent in Human Research." In *Encyclopedia of Bioethics.* 2nd ed. New York: Macmillan, pp. 1241–1250.

Long, Thomas A. and James F. Thorpe. 1987. "The Challenger Case: A Flight From Responsibility?" ASEE Annual Conference Proceedings.

Luegenbiehl, Heinz C. 1983. "Codes of Ethics and the Moral Education of Engineers." In *Business and Professional Ethics Journal,* 2(4):41–61. Reprinted in *Ethical Issues in Engineering,* edited by Deborah Johnson, Englewood Cliffs, NJ: Prentice-Hall, Inc., 1990, pp. 137–154.

Lytton, William B. 1996. "Combating Corruption in Foreign Markets. The Evolving Role of Ethics in Business: Conference Report." The Conference Board, Inc.

MacIntyre, Alasdair. 1981. *After Virtue: A Study in Moral Theory.* Notre Dame, IN: University of Notre Dame Press.

————. 1984a. "Does Applied Ethics Rest on a Mistake?" *The Monist,* 67(4):499–512.

————. 1984b. *Three Rival Versions of Moral Inquiry: Encyclopaedia, Genealogy and Tradition.* Notre Dame, IN: University of Notre Dame Press.

————. 1988. *Whose Justice? Which Rationality?* Notre Dame, IN: University of Notre Dame Press.

Macrina, Francis L. 1995. *Dynamic Issues in Scientific Integrity: Collaborative Research, a Report from the American Academy of Microbiology*. Washington DC: American Academy of Microbiology.

Mann, Charles C. 1994. "Radiation: Balancing the Record." *Science*, 263(5146): 470–474.

Marshall, Eliot. 1996. "Fraud Strikes Top Genome Lab." *Science*, 274 (8 November):908.

Martin, Mike W. and Roland Schinzinger. 19??. *Ethics in Engineering*, second edition. New York: McGraw-Hill Publishing Company.

McConnell, Malcolm. 1987. *Challenger: A Major Malfunction*. Garden City. NY: Doubled.

Mazur, Allan. 1989. "The Experience of Universities in Handling Allegations of Fraud or Misconduct in Research." In *Project on Scientific Fraud and Misconduct: Report on Workshop Number Two*. Rosemary Chalk. Washington DC: American Association for the Advancement of Science, pp. 67–94.

Merchant, Carolyn. 1980. *The Death of Nature: Women, Ecology, and the Scientific Revolution*. San Francisco: Harper & Row.

Middleton, William W. 1986. "Ethical Process Enforcement and Sanctions – The Engineering and Physical Science Societies." Delivered at the AAAS-IIT Workshop on Professional Societies and Professional Ethics, May 23, 1986.

Millikan, Robert A. 1913. "On The Elementary Electrical Charge and the Avogadro Constant." *Physical Review*, 2:109–143.

Mnookin, Seth. "Department of Defense DNA Registry Raises Legal, Ethical Issues." *Gene Watch*, 10(1) (August 1996):1, 3, 11.

Morgenstern, Joe. 1995. "Fifty-Nine Story Crisis." *The New Yorker*, May 29, 1995, pp. 45–53.

Murdoch, Iris. 1969. "On 'God' and 'Good'." In *The Anatomy of Knowledge*, ed. Marjorie Grene. Rutledge and Kegan Paul Ltd. Reprinted in *Revisions: Changing Perspectives in Moral Philosophy*, Stanley Hauerwas and Alasdair MacIntrye, eds., Notre Dame, IN: University of Notre Dame Press, 1983, pp. 68–91.

1971. *The Sovereignty of the Good*. New York: Schocken Books.

Murray, Thomas. 1983. "Warning: Screening Workers for Genetic Risk." *Hastings Center Report*, 13(1):5–8.

1992. "The Human Genome Project and Genetic Testing: Ethical Implications." In *The Genome, Ethics, and the Law*, AAAS-ABA National Conference of Lawyers & Scientists. Washington DC: AAAS.

National Research Council Panel on Scientific Responsibility and the Conduct of Research. 1992. *Responsible Science: Ensuring the Integrity of the Research Process*, Vol. I. Washington DC: National Academy Press.

Nazario, Sonia L. 1991."McDonnell Douglas Jet Evacuation Drills Leave 44 Injured." *Wall Street Journal*, Wednesday, October 30, 1991, A3–A4.

Newhouse, John. 1982. *The Sporty Game*. New York: Alfred A. Knopf, Inc.

Nissenbaum, Helen. 1996. "Accountability in a Computerized Society." *Science and Engineering Ethics*, 2(1).

Nissenbaum, Helen, and Deborah Johnson. 1995. *Computers, Ethics and Social Values*. Englewood Cliffs, NJ: Prentice-Hall.

Norman, Colin. 1988. "Stanford Inquiry Casts Doubt on 11 Papers." *Science*, 242 (4 November 1988):659–661.

Office of the Inspector General, National Science Foundation. 1992. *Semiannual Report to the Congress*. No. 7: April 1, 1992–September 30, 1992, 22.

1993. *Semiannual Report to the Congress*. No. 9: April 1, 1993–September 30, 1993, 37.

PBS. The American Experience: *Rachel Carson's Silent Spring*.

Perrow, Charles.1984. *Normal Accidents*. New York: Basic Books.

Petroski, Henry. 1985. *To Engineer is Human*. New York: St. Martin's Press.

Petulla, Joseph M. 1989. "Environmental Management in Industry." In *Ethics and Risk Management in Engineering*, edited by Albert Flores. Landam, MD: University Press of America.

Pfeifer, Mark P. and Gwendolyn L Snodgrass. 1990. "The Continued Use of Retracted, Invalid Scientific Literature." *Journal of the American Medical Association*, 263(10):1420–23.

Pincoffs, Edmund. 1971. "Quandary Ethics." *Mind*, 80:552–71. Reprinted in *Revisions: Changing Perspectives in Moral Philosophy*, edited by Stanley Hauerwas and Alasdair MacIntrye.

Pooley, Eric. 1996. "Nuclear Warriors." *Time*, March 4, 1996, pp. 46–54.

Postel, Sandra. 1988. "Controlling Toxic Chemicals." In *State of the World*, Linda Starke, ed., pp. 118–136. New York: W. W. Norton & Co.

Potter, Elizabeth. 1994. "Locke's Epistemology and Women's Struggles." In *Modern Engendering*, edited by Bat Ami Bar On. New York: SUNY Press.

Powers, Madison. 1993. "The Right of Privacy Reconsidered." Commissioned paper for *Genetic Privacy Collaboration*, April 22, 1993.

Proctor, Robert. 1988. *Racial Hygiene: Medicine Under the Nazis*. Cambridge, MA: Harvard University Press.

Racker, Efraim. 1989. "A View of Misconduct in Science." *Nature*, 339 (May 1989): 91–93.

Rapp, Rayna. 1987. "Moral Pioneers: Women, Men and Fetuses on a Frontier of Reproductive Technology." *Women and Health*, 13(1–2):101–117. Also published in *Women, Embryos, and Women's Rights: Exploring the New Reproductive Technologies*, edited by Elaine Baruch, Amadeo d'Amado, and Joni Seager.

Rawls, John. 1957. "Outline of Decision Procedure for Ethics." *Philosophical Review*, 66:177–197.

Regan, Tom. 1983. *The Case for Animal Rights*. Berkeley, CA: University of California Press.

Reich, Warren T., editor in chief. 1978. *Encyclopedia of Bioethics*. London: Macmillan the Free Press.

Roberts, Leslie. 1991. "Misconduct: Caltech's Trial by Fire." *Science*, 253:1344–1347.

Roddis, W. M. Kim. 1993. "Structural Failures and Engineering Ethics." *Journal of Structural Engineering ASCE*, (May).

Rorty, Amelie Oksenberg. 1995. "The Many Faces of Morality." *Midwest Studies in Philosophy*, XX:67–82.

Rothstein, Mark A. 1987. "Drug Testing in the Workplace: The Challenge to Employment Relations and Employment Law." *Chicago-Kent Law Review*, 63:683–743.

Rowe, Mary P. 1990. "Barriers to Equality: The Power of Subtle Discrimination to Maintain Unequal Opportunity." *Employee Responsibilities and Rights Journal*, 3(2): 153–163.

Rowe, Mary P. and Michael Baker. 1984. "Are You Hearing Enough Employee Concerns?" *Harvard Business Review*, 62(3):127–135.

Sarasohn, Judy. 1993. *Science on Trial*. New York: St. Martin's Press.

Scanlon, Walter E. 1980. *Alcoholism and Drug Abuse in the Workplace: Employee Assistance Programs*. New York: Prager.

Schneewind, J. B. 1985. "Applied Ethics and the Sociology of the Humanities." In *Applying the Humanities*, edited by Denial Callahan, Arthur L. Caplon, and Bruce Jennings, pp. 57–70. New York: Plenum Press.

"Moral Knowledge and Moral Principles." 1968–69. *Knowledge and Necessity, Royal Institute of Philosophy Lectures*, 3:249–62, reprinted in *Revisions: Changing Perspectives in Moral Philosophy*, edited by Stanley Hauerwas and Alasdair MacIntrye.

Schneider, Keith. 1991. "Military Has New Strategic Goal In Cleanup of Vast Toxic Waste." New York Times, August 5, 1991, A1, D3.

Schwab, Adolf J. 1996. "Engineering Ethics in the U.S. and Germany." *IEEE Institute*, June, 1996.

Senge, Peter M. 1990. *The Fifth Discipline, The Art & Practice of the Learning Organization*. New York: Doubleday.

Shore, Eleanor. 1995. "Effectiveness of Research Guidelines in Prevention of Scientific Misconduct." *Science and Engineering Ethics*, 1(4) (October):383.

Singer, Peter. 1977. *Animal Liberation*. New York: Avon Books.

1985. *In Defense of Animals 1985*. New York: Blackwell.

1990. "The Significance of Animal Suffering." *Behavioral and Brain Sciences*, 13(1):9–12.

Stewart, Walter and Ned Stewart. 1987. "The Integrity of the Scientific Literature." *Nature*, 325 (January 15):207–214.

Susskind, Lawrence E. 1990. "A Negotiation Credo for Controversial Siting Disputes." *Negotiation Journal*. October.

Tannenbaum Jerrold and Andrew Rowan. 1985. "Rethinking the Morality of Animal Research." *Hastings Center Report*, 15 October:32–36.

Taubes, Gary. 1993. *Bad Science*. New York: Random House.

Thomson, Judith Jarvis. 1990. *The Realm of Rights*. Cambridge MA: Harvard University Press.

Thoreau, Henry David. "Civil Disobedience." Reprinted in *Civil Disobedience in Focus*, pp. 28–48, edited by Hugo A. Bedau. New York and London: Routledge, 1991.

Thorpe, James F., and William H. Middendorf. 1979. *What Every Engineer Should Know About Product Liability*. New York and Basal: Marcel Dekker, Inc.

Toulmin, Stephen. 1981. "The Tyranny of Principles: Regaining the Ethics of Discretion." *Hastings Center Report*, 11(6):31–39.

Turkington, Richard, George, Trubow, and Anita, Allen. 1992. *Privacy: Cases and Materials*. Huston: John Marshall Press.

Unger, Stephen H. 1994. *Controlling Technology: Ethics and the Responsible Engineer*. second edition. New York: John Wiley.

Vesilind, P. Aarne, J. Jeffrey Peirce, and Ruth Weiner. 1987. *Environmental Engineering*. 2nd edition. Stoneham, MA: Butterworth.

Wald, Mathew L. 1996. "Two Northeast Utilities Plants Face Shutdown." *The New York Times*, March 9, 1996.

Weil, Vivian. 1983a. "Beyond Whistleblowing: Defining Engineers' Responsibilities." *Proceedings of the Second National Conference on Ethics in Engineering*. Center for the Study of Ethics in the Professions, Illinois Institute of Technology, Chicago, IL.

1983b. "The Browns Ferry Case." In *Engineering Professionalism and Ethics*, James H. Schaub and Karl Pavlovic, eds. New York: John Wiley & Sons.

1989. "Military Research, Secrecy, and Ethics." In *Ethical Issues Associated with Scientific and Technological Research for the Military*, Carl Mitcham and Philip Siekevitz, eds. pp. 193–99. *Annals of the New York Academy of Sciences*, vol. 577.

Weil, Vivian and Rachelle Hollander. 1990. "Sharing Scientific Data II: Normative Issues." *IRB*, 12:7–8.

Weil, Vivian and John W. Snapper eds. 1989. *Owning Scientific and Technical Information: Value and Ethical Issues*. New Brunswick, NJ: Rutgers University Press.

Weinstein, Milton C. and Harvey V. Fineberg 1980.*Clinical Decision Analysis*. W. B. Saunders Company.

Weiss, Philip. 1989. "Conduct Unbecoming?" *The New York Times Magazine*, October 29, pp. 40–41, 68.

Weisskoph, Steven. 1989. "The Aberdeen Mess." *Washington Post Magazine*, January 15, p. 55.

Westin, Alan F. 1988. *Resolving Employment Disputes Without Litigation*. Washington, DC: Bureau of National Affairs.

Westrum, Ron. 1991. *Technologies and Society*. Belmont, CA: Wadsworth. Chapter 7: The Role of Designers in Technology.

Wheeler, David L. 1992. "U.S. Attorney: Leave 'Baltimore Case' to the Scientists." *The Chronicle of Higher Education*, July 22, 1992. A7.

1995. "Making Amends to Radiation Victims." *The Chronicle of Higher Education*, October 13, 1995.

Whitbeck, Caroline. 1985. "Why the Attention to Paternalism in Medical Ethics?" *Journal of Health Politics, Policy and Law*, 10(1):181–187.

1991. "Ethical Issues Raised by New Medical Technologies." In *The New Reproductive Technologies*, pp. 49–64. Hillsdale, NJ: Lawrence Erlbaum Publishers.

1992a. "Ethical Considerations in Biomonitoring Communities for Toxic Contamination." *Technology and Discovery: Proceedings of the Sixth International Conference of Society for Philosophy and Technology*, edited by Joe Pitt and Elana Lugo and slightly adapted from the author's contribution to *Monitoring the Community for Exposure and Disease*, a report to the Agency for Toxic Substances and Disease Registry. Nicholas Ashford, Principal Investigator.

1992b. "The Trouble With Dilemmas: Rethinking Applied Ethics." *Professional Ethics*, 1(1 & 2):119–142.

1995. "Trust." In *Encyclopedia of Bioethics*, second edition. New York: Macmillan, pp. 2499–2504.

Williams, Bernard. 1985. *Ethics and the Limits of Philosophy*. Cambridge, MA: Harvard University Press.

1988. "Formal Structures and Social Reality." In *Trust: Making and Breaking Cooperative Relations*. edited by Diego Gambetta. Oxford: Basil Blackwell, pp. 3–13.

1993. *Shame and Necessity*. Berkeley: University of California Press.

Winslow, Ron. 1994. "FDA Halts Tests on Device That Shows Promise for Victims of Cardiac Arrest." *Wall Street Journal*, May 11, 1994, p. B8.

Woolf, Patricia K. 1986. "Pressure to Publish and Fraud in Science." *Annals of Internal Medicine*, 104(2):254–256.

Worobec, M. 1980. "An Analysis of the Resource Conservation and Recovery Act." *BNA Government Reporter*, Special Report (August 22).

Young, Iris. 1990. *Justice and the Politics of Difference.* Princeton: Princeton University Press.

Zuckerman, Diana M. 1993. "Conflict of Interest and Science." *The Chronicle of Higher Education*, October 13, 1993, B1.

Zuras, A. D., F. J. Prinznar, and C. S. Parrish. 1985. "The National Priorities List Process." In *Management of Uncontrolled Hazardous Waste Sites*, ed. AIChE. New York: AIChE, p. 1–3.

INDEX

Boldface page numbers indicate definitions; italic page numbers indicate case studies.

Petrasso, Richard, 220
Petroski, Henry, 108, 117, 196, 213
Petulla, Joseph M., 258
Physical examinations, preemployment,
    174
plagiarism, 49, 69–70, 200, **201**, 218, 223
    n.13, 264–5, 268–71, 283–4, 291, 292,
    302
Poincare, Henri, 31–2
Pons, Stanley, 216, 219–20
Popovic, Mikulus, 226 n.55
Powers, Madison, 47
preferences, 3–6, 17
    investigator's, 208
prejudice, 159, 182, 185–6, 234
    see also discrimination
privacy, 2–3, **46**–7, 170–5
Proctor, Robert, 35
product testing, 231
professions, 40, 74–82, 99, 104, 112, 217
    see also engineering profession
    autonomy of, 93–100, **94**, 104, 217
professional societies, 74–6, 93, 94–6, 102
    codes of ethics, 28, 29–30, 38, 43, 48, 75,
        76–82, 93–100, 104, 105, 110, 182,
        244, 281, 298
    election to, 267
    employment guidelines, 168–70, 190
    publication guidelines, 278–84
property, 200
    see also intellectual property; rights,
        property
protest, 50, 84–5, 101–2, 238–9
    see also complaint
Public Health Service, United States, 204,
    226 n.55, 227

Racker, Efraim, 207, 222
radiation experiments, 228
reasoning, 208–9
recklessness, 70, 196, 216, 217, 218, 220,
    221–2, 255
recommendation, letters of, 93
Regan, Tom, 233
relativism
    cultural, 10–1
    ethical, 10–1, 58
religion and religious standards, 8
reparations, 169
replication of research results, 196–7
research integrity, 194–226, 282
research misconduct, 62–3, 162–3, 194, 195,
    196–7, 198, **200**–3, 205, 206, 213, 214,
    218, 219–21, 222, 226 n.55, 264, 269,
    270, 275, 291
Research Triangle Institute, 171
Resource Conservation and Recovery Act,
    91–2, 240

responsibility
    environmental, 236–62
    ethical, 18, 21, **37**–44, 57–8, 75, 83, 112,
        167, 181, 200, 215, 243, 245, 247
    legal, 41
    official, 28, **40**–2, 245–6
    professional, **39**–40, 42–3, 44, 57–8, 70–1
        71–2, 81, 74–108, 109–32, 167, 196,
        208, 215, 222–3 n.4, 243, 247, 259
    trust and, 45–6, 167–8
retractions, 99, 274
reviewers, responsibilities of, 99, 196, 206,
    209, 217, 218–9, 278–9, 281, 283–4,
    284–5, 286, 291
Ride, Sally, 143
right to life, 19, 21, 27, 29
rights
    absolute, **24**–5, 27, 28
    alienable/inalienable, 21–2, 23, 24–5, 27,
        48
    civil, **19**, 182–5
    economic, **26**
    ethical, 18–27, 49, 82–3
    exercise of, **22**–3
    human 2, 18, 19, 20, 21, 26, 27, 34–5, 39,
        46, 60, 84–6, 248
    infringement of, **24**, 43, 297
    legal, **19**, 32, 49–50, 247–8, 271, 298–9, 301
    natural, **19**, 27
    negative, **25**, 26, 27, 29, 32
    political, 19, **26**
    positive, **25**–6, 27, 29, 32
    prima facie, **24**, 25, 27, 28
    property or proprietary, 19, 22, 24, 25, 27,
        45–6, 48, 169–70, 271, 297–9
    special, **18**, 48
    violation of, **24**, 60, 84–6
    waiver of, **22**–3
risk, **16**, 119, 119, 121, 141, 218, 229
    assessment, 16–7, 18, 113, 242, 244, 245–6
    management, 242
    shifting, **18**
Roosevelt, Eleanor, 19
Roosevelt, Theodore, 238
Rorty, Amélie Oskenberg, 9, 13
Rorty, Richard, 13
Roth, Jesse, 284
Rowe, Mary, 185
Rowland, Sherwood, 254

sabotage of experiments, 202
safety
    complaint procedures, 42, 141, 143, 161,
        166
    drug testing of employees and, 171
    engineer-manager communications on,
        157–60
    hazards, 108, 121, 239, 243, 283